electricity four

Hayden Electricity One—Seven Series

Harry Mileaf, Editor-in-Chief

electricity one Producing Electricity □ Atomic Theory □ Electrical Charges □ Electron Theory □ Current □ Voltage □ Magnetism □ Electromagnetism

electricity two D-C Circuits □ Direct Current □ Resistors □ Ohm's Law □ Power □ Series Circuits □ Parallel Circuits □ Series-Parallel Circuits □ Kirchhoff's Laws □ Superposition □ Thevenin's Theorem □ Norton's Theorem

electricity three A-C Circuits □ Alternating Current □ A-C Waveforms □ Resistive Circuits □ Inductors □ Inductive Circuits □ Transformers □ Capacitors □ Capacitive Circuits

electricity four LCR Circuits □ Vectors □ RL Circuits □ RC Circuits □ LC Circuits □ Series-Parallel Circuits □ Resonant Circuits □ Filters

electricity five Test Equipment □ Meter Movements □ Ammeters □ Voltmeters □ Ohmmeters □ Wattmeters □ Multimeters □ Vacuum-Tube Voltmeters

electricity six Power Sources □ Primary Cells □ Batteries □ Photo, Thermo, Solar Cells □ D-C Generators □ A-C Generators □ Motor-Generators □ Dynamotors

electricity seven Electric Motors □ D-C Motors □ A-C Motors □ Synchronous Motors □ Induction Motors □ Reluctance Motors □ Hysteresis Motors □ Repulsion Motors □ Universal Motors □ Starters □ Controllers

electricity four

HARRY MILEAF EDITOR-IN-CHIEF

HAYDEN BOOKS
A Division of Howard W. Sams & Company
4300 West 62nd Street
Indianapolis, Indiana 46268 USA

SECOND EDITION
TENTH PRINTING—1989

International Standard Book Number: *0-8104-5948-5*
Library of Congress Catalog Card Number: *75-45504*

Printed in the United States of America

preface

This volume is one of a series designed specifically to teach electricity. The series is logically organized to fit the learning process. Each volume covers a given area of knowledge, which in itself is complete, but also prepares the student for the ensuing volumes. Within each volume, the topics are taught in incremental steps and each topic treatment prepares the student for the next topic. Only *one* discrete topic or concept is examined on a page, and *each* page carries an illustration that graphically depicts the topic being covered. As a result of this treatment, neither the text nor the illustrations are relied on solely as a teaching medium for any given topic. Both are given for *every* topic, so that the illustrations not only complement but reinforce the text. In addition, to further aid the student in retaining what he has learned, the important points are summarized in text form on the illustration. This unique treatment allows the book to be used as a convenient review text. Color is used not for decorative purposes, but to accent important points and make the illustrations meaningful.

In keeping with good teaching practice, all technical terms are defined at their point of introduction so that the student can proceed with confidence. And, to facilitate matters for both the student and the teacher, key words for each topic are made conspicuous by the use of italics. Major points covered in prior topics are often reiterated in later topics for purposes of retention. This allows not only the smooth transition from topic to topic, but the reinforcement of prior knowledge just before the declining point of one's memory curve. At the end of each group of topics comprising a lesson, a summary of the facts is given, together with an appropriate set of review questions, so that the student himself can determine how well he is learning as he proceeds through the book.

Much of the credit for the development of this series belongs to various members of the excellent team of authors, editors, and technical consultants assembled by the publisher. Special acknowledgment of the contributions of the following individuals is most appropriate: Frank T. Egan, Jack Greenfield, and Warren W. Yates, principal contributors; Peter J. Zurita, Steven Barbash, Solomon Flam, and A. Victor Schwarz, of the publisher's staff; Paul J. Barotta, Director of the Union Technical Institute; Albert J. Marcarelli, Technical Director of the Connecticut School of Electronics; Howard Bierman, Editor of *Electronic Design;* E. E. Grazda, Editorial Director of *Electronic Design;* and Irving Lopatin, Editorial Director of the Hayden Book Companies.

HARRY MILEAF
Editor-in-Chief

contents

This volume describes circuits that contain:

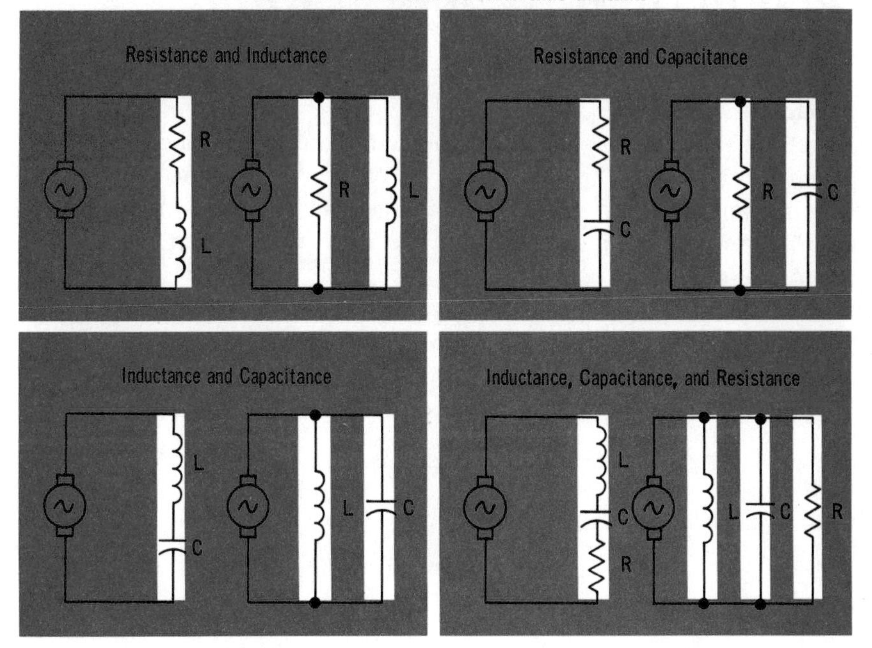

introduction

In Volume 3, you learned the properties of alternating current and how it is used. You also learned about inductance and capacitance, and were shown how to analyze a-c circuits that contain resistance, inductance, or capacitance. Volume 3 was limited, however, to circuits having only *one* of these circuit elements. The operation of circuits having two of the elements, such as resistance and capacitance or resistance and inductance, or even all three of the elements, was not covered. These more complex circuits are the major subject of this volume. You will learn how to analyze and solve circuits that contain both series and parallel combinations of resistance and inductance; resistance and capacitance; inductance and capacitance; and inductance, capacitance, and resistance. You will also learn some of the unique applications of these circuits.

Much of the material in this volume is based on the *phase relationships* between the voltages and currents in the various circuits. These relationships can be both explained and understood more easily if *vectors* are used to describe them. Therefore, before the actual circuits are described, a basic description of vectors and their use in electricity will be given.

review of LCR phase angles

You will remember from Volume 3 that the term *phase angle* is used to describe the *time* relationship between a-c voltages and currents, as well as to specify a *position* or *point* in time of one a-c voltage or current. For example, if two a-c voltages are of opposite polarity at every instant of time, they are 180 degrees out of phase, or the phase angle between them is 180 degrees. Similarly, if a current reaches its maximum amplitude after one-quarter of its cycle, or 90 degrees, maximum amplitude is said to occur at a phase angle of 90 degrees.

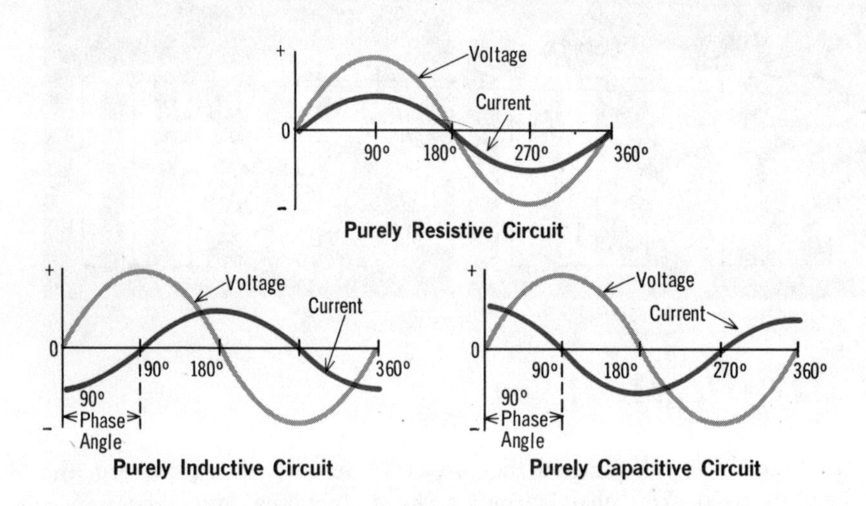

In a purely resistive circuit, the phase angle between the voltage and current is zero. In a purely inductive circuit, the phase angle is 90°, with the voltage leading. In a purely capacitive circuit, the phase angle is 90°, with the current leading

You also learned in Volume 3 that there are very definite phase relationships between the *applied voltage* and the *circuit current* in purely resistive, purely inductive, and purely capacitive circuits. These relationships can be summarized as follows:

1. In a purely *resistive* circuit, the voltage and current are *in phase.*
2. In a purely *inductive* circuit, the applied *voltage leads* the *current* by 90 degrees.
3. In a purely *capacitive* circuit, the *current leads* the applied *voltage* by 90 degrees.

You already know how these relationships can be described using voltage and current *waveforms*. However, there is another, and often easier, way to express these same relationships. This is by using *vectors*.

what is a vector?

Every physical quantity has *magnitude.* The terms "5 apples," "5 days," and "5 ohms" all express physical quantities, and each is *completely* described by the number 5. There are some quantities, however, that are *not* completely described if only their magnitudes are given. Such quantities have a *direction* as well as magnitude, and without the direction these quantities are *meaningless.* For example, if

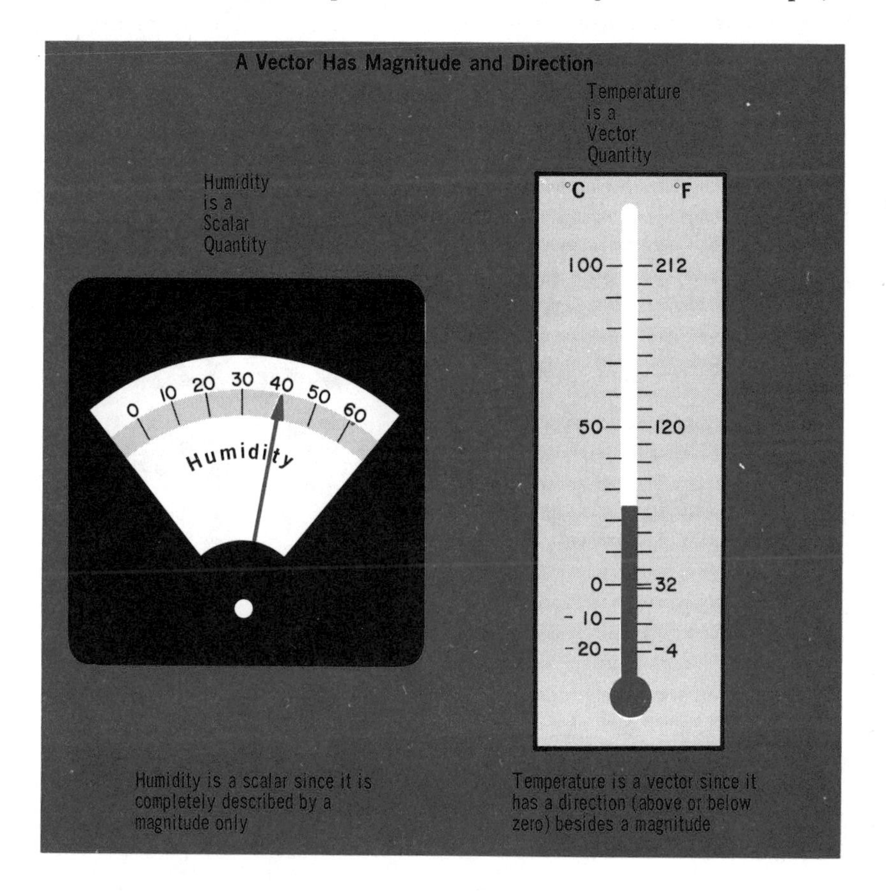

someone were to ask you how to get from Chicago to Milwaukee, and you told them to drive 80 miles, this would be meaningless. But if you said to drive 80 miles *north,* your directions would be complete. The quantity "80 miles north" thus has a magnitude of 80 and a direction of north.

Quantities that have only magnitude are called *scalars.* Those that have magnitude and direction are called *vectors.*

graphical representation

Graphically, a vector is represented by a *straight line* with an *arrowhead* at one end. The *length* of the line is proportional to the *magnitude* of the vector quantity, and the *arrowhead* indicates the *direction*. When a vector is drawn on a graph, the direction of the vector is usually represented by the *angle* it makes with the *horizontal* axis. The horizontal axis thus serves as the *reference line* from which the direction of the vector is measured. You will learn later that in actual practice this reference point can be a direction, such as north, or a position in time, such as the zero-degree point of a sine wave.

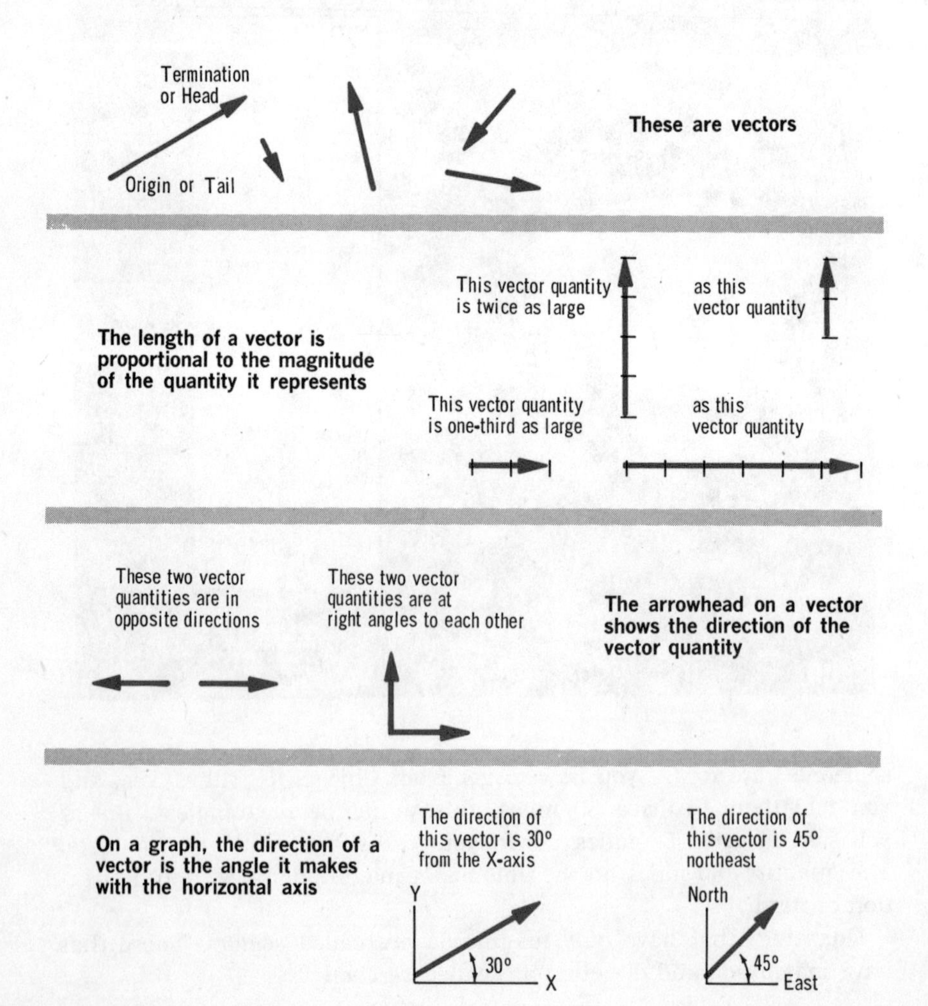

working with vectors

Since vectors represent physical quantities, they can be added, subtracted, multiplied, and divided the same as other physical quantities. In your work with electricity, you will normally only be interested in the *addition* and *subtraction* of vectors. Because of this, vector multiplication and division will not be covered in this volume.

Vector Quantities Can Be Added and Subtracted

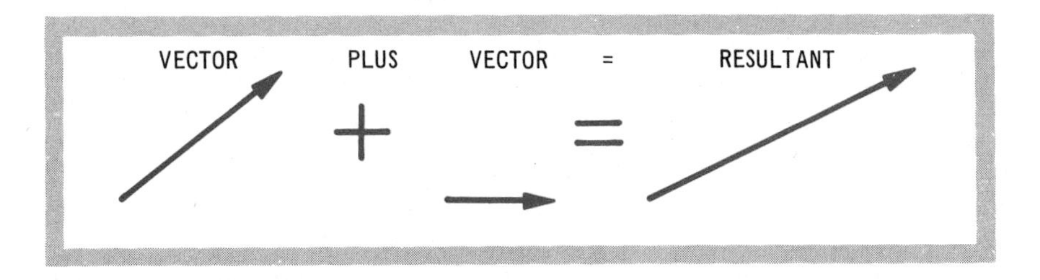

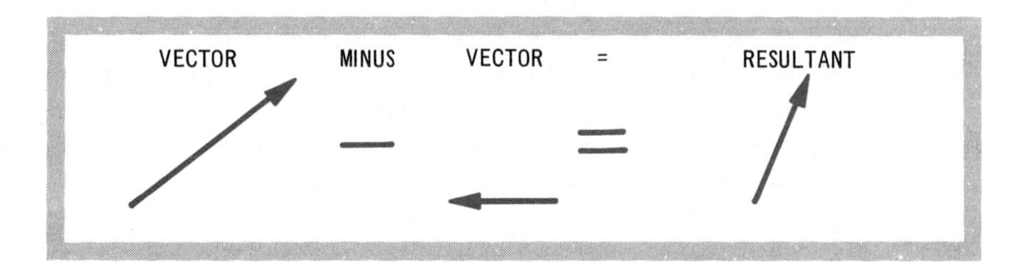

When vectors are added or subtracted, the result is also a vector, and is called the resultant vector, or just a resultant

The methods used to add and subtract scalar quantities *cannot* be used for vector quantities. The fact that vectors have direction, as well as magnitude, makes it necessary to add and subtract them *geometrically*. As a result, the adding and subtracting of vectors is a combination of geometry and *algebraic* addition and subtraction.

There are various methods you can use when adding or subtracting vectors. The particular method you select will depend on the relative directions of the vectors involved.

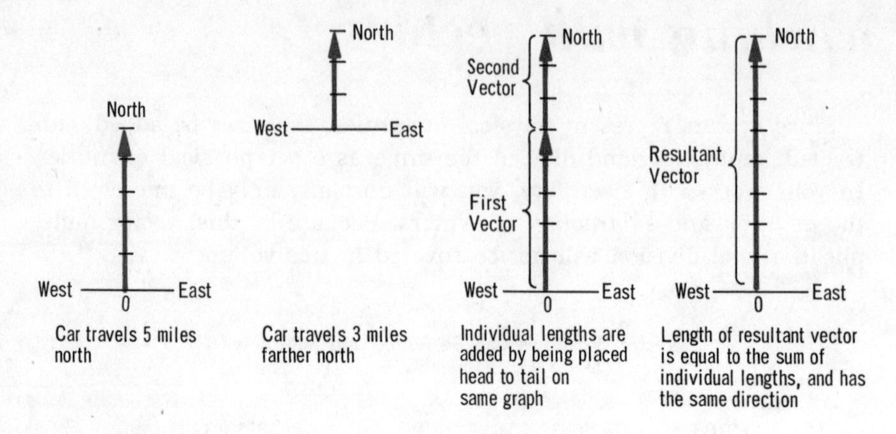

adding vectors that have the same direction

To add vector quantities that have the *same* direction, you simply *add* the individual *magnitudes*. This gives you the magnitude of the resultant. Its direction is the *same* as the direction of the individual vectors. An example of this would be a car that traveled 5 miles north from a starting point. This motion could be represented by a vector pointing north, and having a length corresponding to 5 miles. If the car then traveled 3 miles farther, still going north, the additional motion could be represented by another vector. This vector would also point north, but would have a length corresponding to 3 miles. The total motion of the car, or its final position, can be found by adding the lengths of the two vectors and assigning the resultant a direction of north. The resultant then shows that the car has traveled a total of 8 miles north of the starting point.

Series Aiding Batteries is an Example of Adding Vectors Having the Same Direction

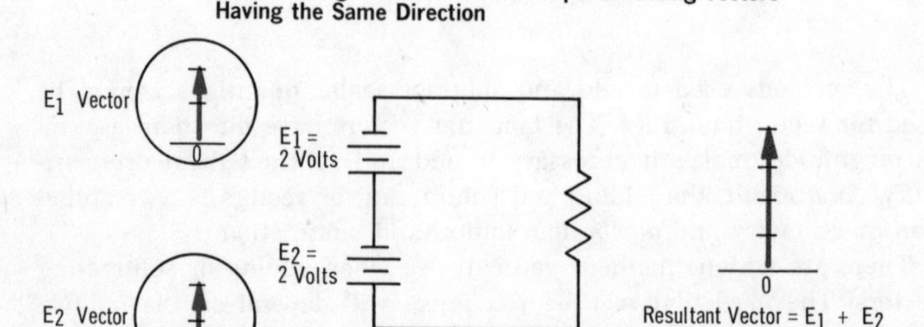

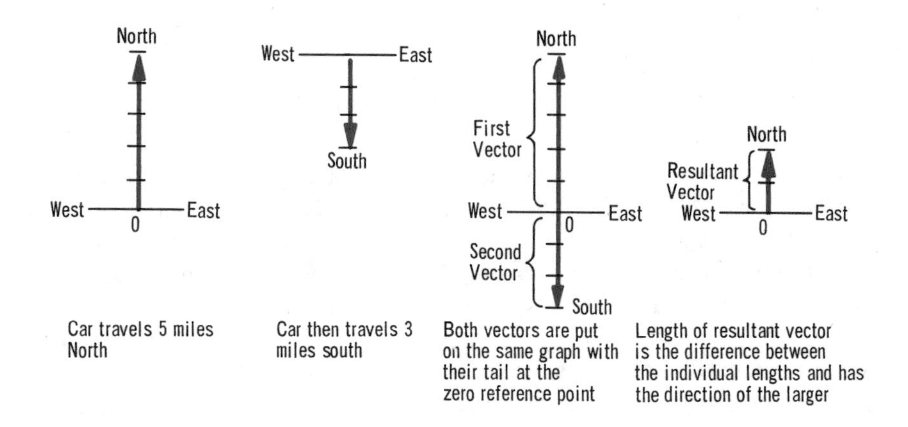

Car travels 5 miles North

Car then travels 3 miles south

Both vectors are put on the same graph with their tail at the zero reference point

Length of resultant vector is the difference between the individual lengths and has the direction of the larger

adding vectors that have opposite directions

To add vector quantities that have *opposite* directions, *subtract* the one with the *smaller* magnitude from the one with the *larger* magnitude. This gives the magnitude of the resultant. The *direction* of the resultant is the same as that of the *larger* vector. You can see from this that if one vector is added to another that has the same magnitude but opposite direction, the resultant is zero.

An example of the addition of vectors that have opposite directions is shown. If a car travels north for 5 miles, the vector representing the motion points north, and has a length corresponding to 5 miles. If the car then travels 3 miles south, the vector representing this additional motion points south, and has a length corresponding to 3 miles. The resultant of these two motions is found by adding the two vectors. Since their directions are opposite, this is done by subtracting the smaller vector (3 miles) from the larger vector (5 miles), and assigning the resultant the direction of the larger vector (north). The resultant then shows that after the two motions, the car is at a position 2 miles north of the starting point.

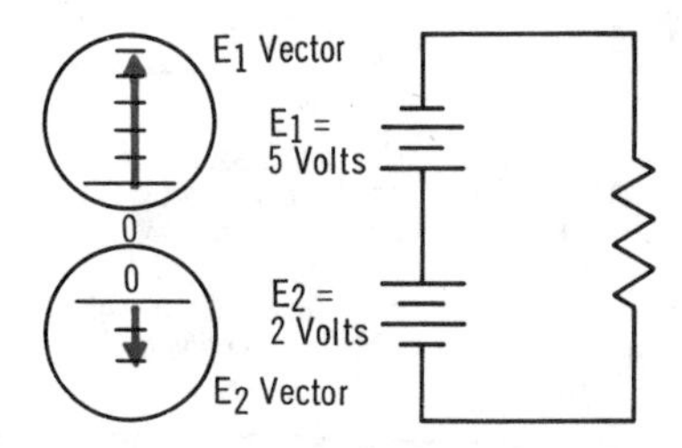

Series Opposing Batteries is an Example of Adding Vectors That Have Opposite Directions

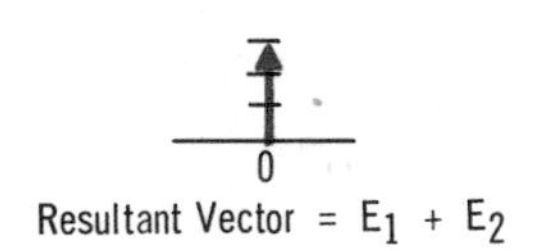

adding vectors by parallelogram method

When vectors are neither in the same nor in opposite directions they can be added *graphically* and their resultant found by means of the *vector parallelogram method.* You will remember from mathematics that a parallelogram is a *four-sided* figure whose opposite sides are *equal* in length and *parallel* to each other. To add two vectors using the parallelogram method, the vectors must first be placed "tail to tail." A parallelogram, then, is constructed by using the vectors. The *diagonal* of a parallelogram so constructed from the tail of the vectors is the *resultant* of the two vectors. The length of the diagonal is the length of the resultant, and the angle between the diagonal and the horizontal axis (θ) represents the direction of the resultant.

The parallelogram method can be used to add more than two vectors. To do this, you first find the resultant of any two of the vectors. You then use this resultant with one of the remaining vectors, and find the resultant of that combination. This resultant is then used with another vector, and so on, until the overall resultant has been found for all the vectors.

You can see that the parallelogram method is primarily a graphical method for adding vectors. It requires the use of a ruler and a protractor, and at best is only fairly accurate, unless great care is taken. For this reason, you will find that you very rarely use the parallelogram method to find numerical solutions to vector addition problems when working with electricity. It is nevertheless important to understand, since it is often used to analyze and describe relationships between vector quantities.

Constructing a Vector Parallelogram

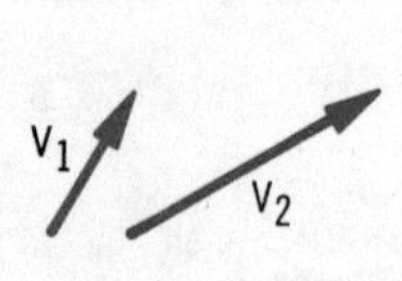

(A) To add two vectors

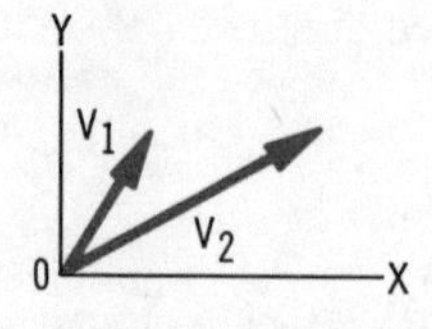

(B) First place them tail to tail

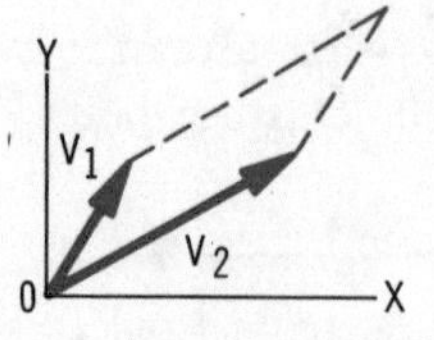

(C) Then construct a parallelogram using the vectors as two of the sides

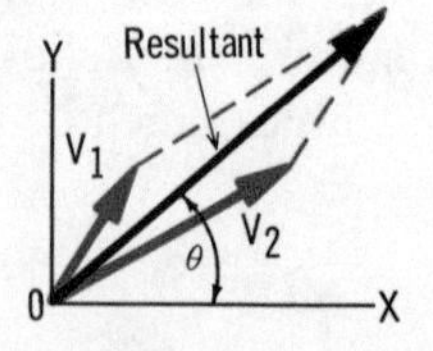

(D) The diagonal of the parallelogram is the resultant. The length of the diagonal is the magnitude and the angle θ is the direction of the resultant

adding vectors by triangle method

Another method for *graphically* adding vectors is called the *vector triangle method.* As its name implies, the triangle method involves the construction of a triangle to determine the resultant vector. To use the method for adding two vectors, the vectors are first placed "head to tail." A line is then drawn connecting the *tail* of the *first* vector with the *head* of the *second.* This is the resultant of the two vectors.

Constructing a Vector Triangle

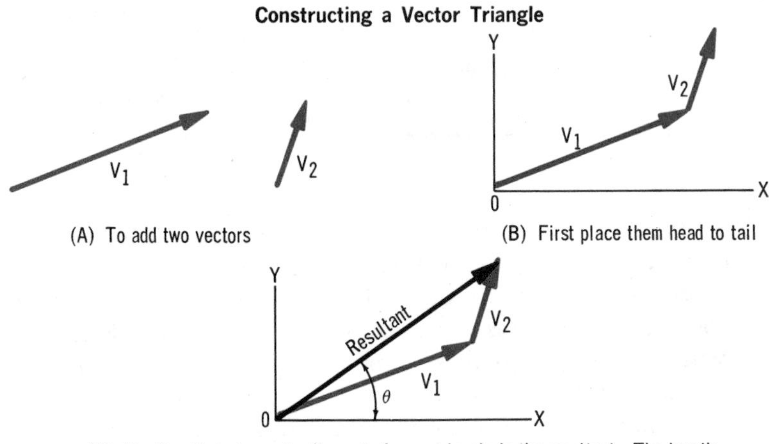

(C) The line that closes the figure to form a triangle is the resultant. The length of the line is the magnitude, and the angle θ is the direction of the resultant

The triangle method can be used when there are more than two vectors involved. To do this, all of the vectors are connected "head to tail." A line is then drawn from the tail of the first vector to the head of the last vector. This line is the total resultant of all the individual vectors.

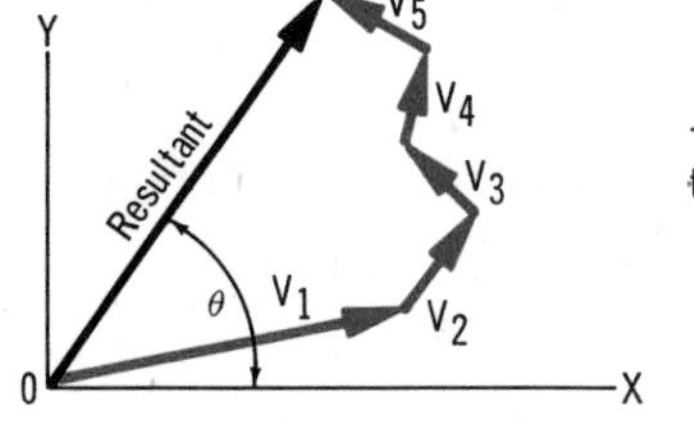

The triangle method can be used to find the resultant of more than two vectors

Like the parallelogram method, the triangle method is used mainly to analyze and describe relationships between vector quantities, rather than to obtain numerical solutions to vector additions.

adding vectors that are 90 degrees apart

A very special case of vectors that are neither in the same nor in opposite directions are those that are 90 *degrees apart.* As you will learn later, these are especially important in electricity because of their relationship to the 90-degree phase difference that exists between the voltage and current in inductive and capacitive circuits.

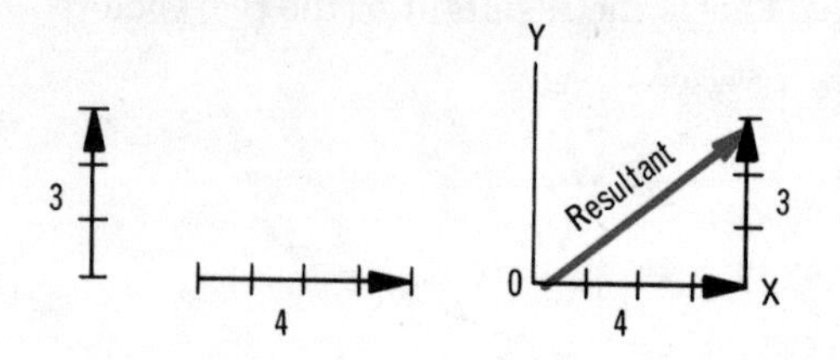

Two vectors that are 90° apart can be added graphically using the vector triangle method. The resultant and the two vectors form a right triangle, so the magnitude of the resultant can be calculated by the Pythagorean Theorem

Graphically, two vectors that are 90 degrees apart can be added using either the parallelogram method or the triangle method. When vector addition is done, the resultant is the *hypotenuse* of a *right triangle* whose other two sides are the vectors being added. Because of the properties of the right triangle, summarized on pages 4-11 and 4-12, the magnitude and direction of the resultant can easily be calculated. As shown if the magnitudes of two vectors are known, and the vectors are 90 degrees apart, the magnitude of the resultant can be calculated by the *Pythagorean Theorem* ($c^2 = a^2 + b^2$). The direction of the resultant, which is the angle (θ) between it and the horizontal axis, can then be found using the *trigonometric relationship* between the sine of the angle and the sides of the triangle.

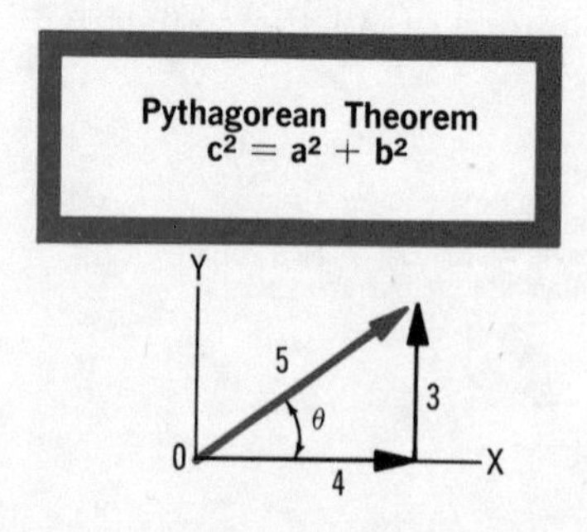

$$(\text{Resultant})^2 = a^2 + b^2$$

$$\text{Resultant} = \sqrt{a^2 + b^2}$$

$$= \sqrt{9 + 16} = \sqrt{25} = 5$$

The direction of the resultant can be found using the trigonometric relationship:

$$\sin\theta = \frac{\text{opposite}}{\text{hypotenuse}} = \frac{3}{5} = 0.6$$

From a table of trigonometric functions, therefore: $\theta = 36.9°$

It might be advisable at this time for you to review right triangles and their solution. Such material is available in any standard trigonometry text, or in books covering mathematics for electricity.

properties of the right triangle

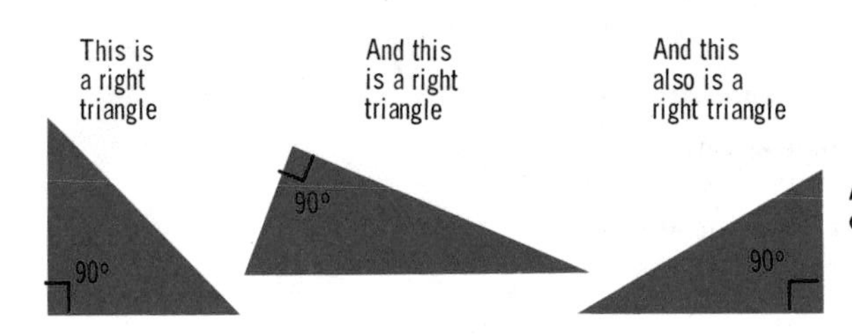

A right triangle is one that contains a 90° angle

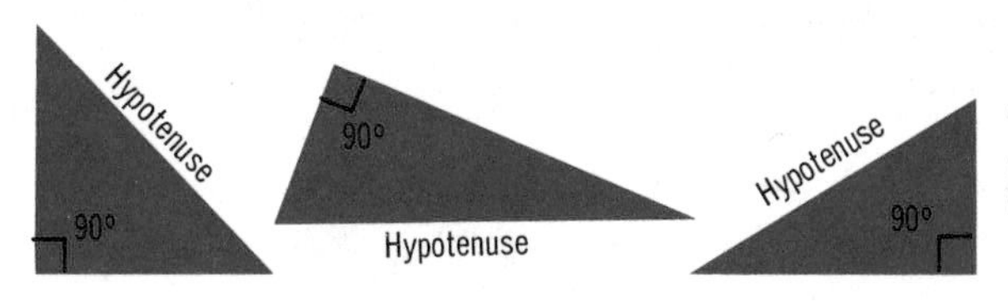

The side of the triangle opposite the 90° angle is called the hypotenuse

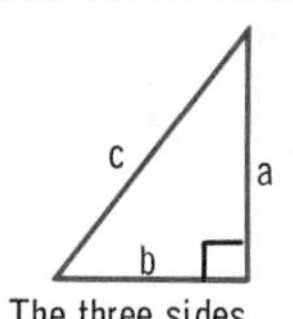

The three sides are designated a, b, and c, with c being the hypotenuse

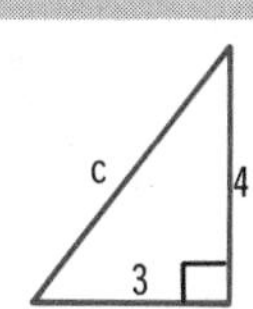

If any two sides are known, the third side can be found by the Pythagorean Theorem

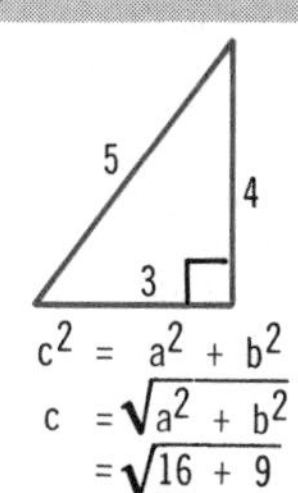

$$c^2 = a^2 + b^2$$
$$c = \sqrt{a^2 + b^2}$$
$$= \sqrt{16 + 9}$$
$$= \sqrt{25}$$
$$= 5$$

The lengths of the three sides of a right triangle are related by the Pythagorean Theorem:

$$c^2 = a^2 + b^2$$

where c is the length of the hypotenuse, and a and b are the lengths of the other two sides

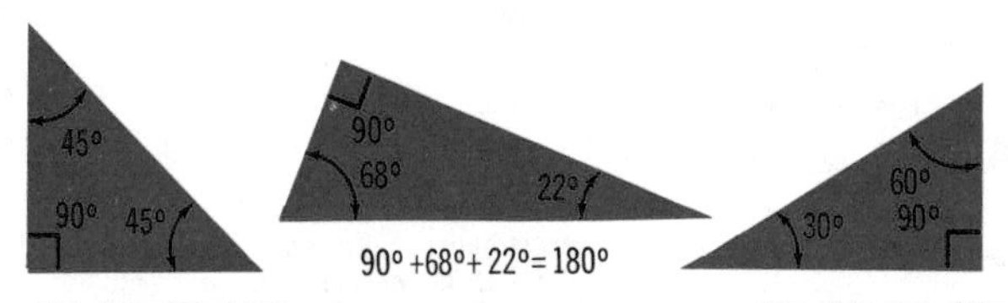

$90° + 45° + 45° = 180°$

$90° + 68° + 22° = 180°$

$90° + 30° + 60° = 180°$

The sum of the three angles in any right triangle is 180°

The properties of the right triangle are continued on the next page.

properties of the right triangle (cont.)

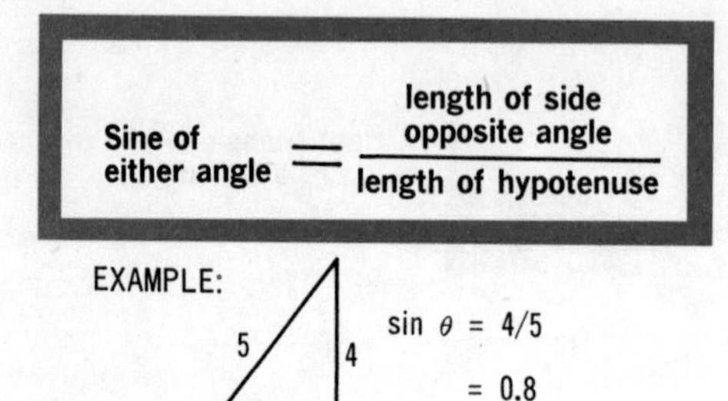

The two angles of a right triangle smaller than 90° are related to the lengths of the sides by the trigonometric relationships of sine, cosine, and tangent

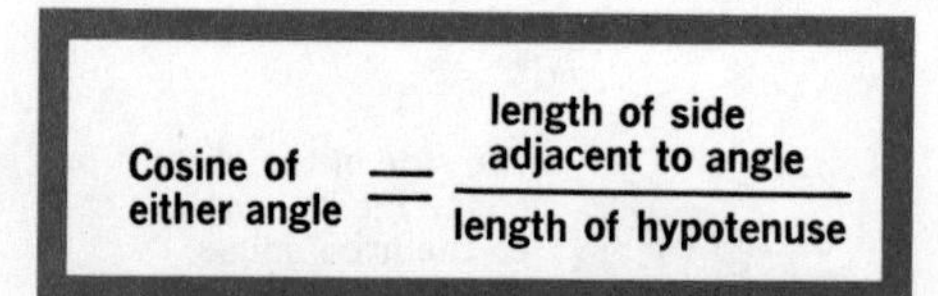

EXAMPLE:

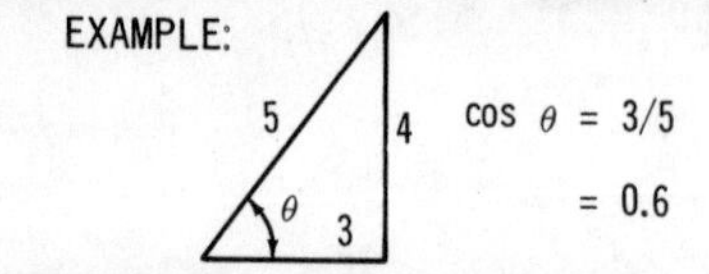

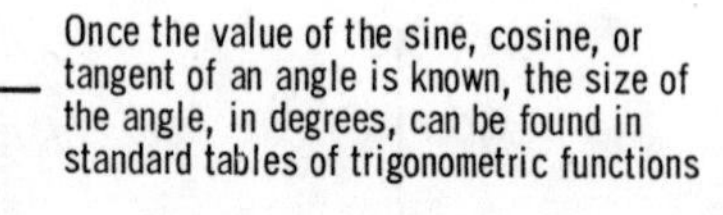
Once the value of the sine, cosine, or tangent of an angle is known, the size of the angle, in degrees, can be found in standard tables of trigonometric functions

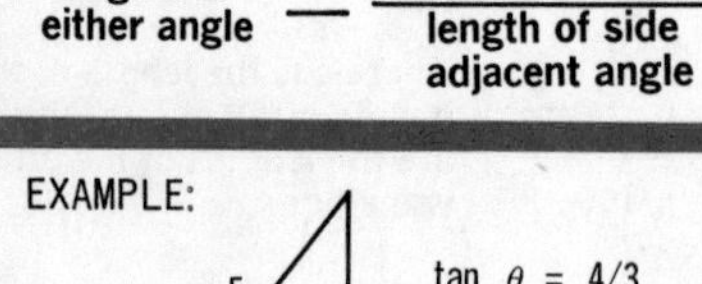

EXAMPLE:

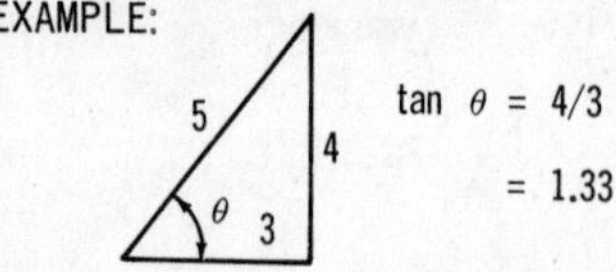

Since $\cos\theta = a/c$

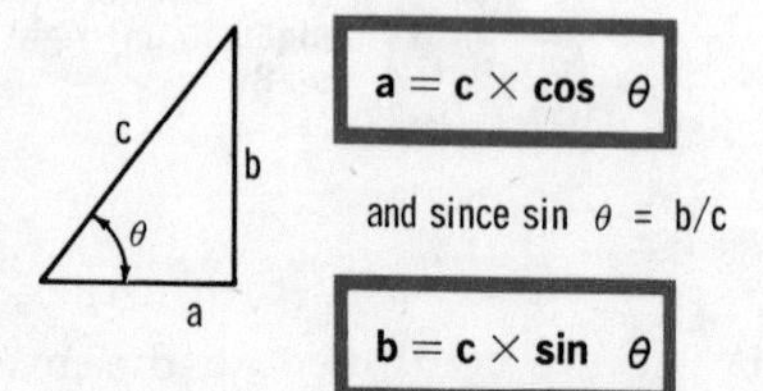

If the length of the hypotenuse and the angle between it and one of the other sides is known, the length of the other two sides can be found using the trigonometric relationships of sine or cosine

subtracting vectors

To subtract one vector from another, *add* the *negative equivalent* of the vector to be subtracted to the other vector. This means that once the vector to be subtracted has been converted to its negative equivalent, the two vectors are merely added using whichever of the methods of *vector addition* that is appropriate.

The negative equivalent of a vector is obtained by *rotating* the vector *180 degrees*. The magnitude of the vector is thus the same, but its direction is reversed.

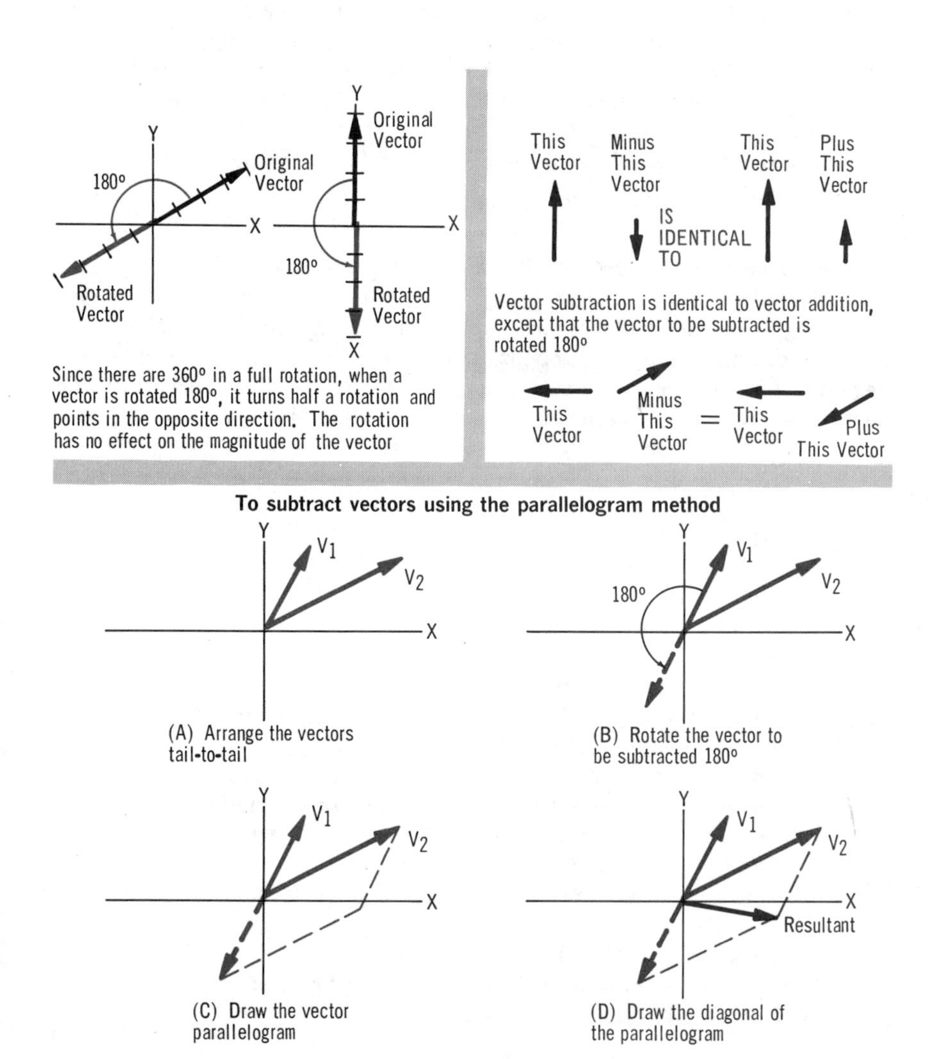

summary

☐ Phase angles are used to describe the time relationships between a-c voltages and currents. They also specify the position or point in time of one voltage or current. ☐ In a purely resistive circuit, the voltage and current are in phase. ☐ In a purely inductive circuit, the applied voltage leads the current by 90 degrees. ☐ In a purely capacitive circuit, the current leads the applied voltage by 90 degrees. ☐ Quantities that have magnitude only are called scalars; those having both magnitude and direction are called vectors. ☐ Vectors are often used to express voltage and current relationships.

☐ Vectors that have the same direction can be added by simply adding the magnitude of the individual vectors. ☐ The sum of the vectors that have opposite directions is a vector whose magnitude is the difference of the two vectors, and whose direction is the same as that of the larger vector. ☐ The parallelogram method is used, in general, to graphically add two vectors. ☐ Another method for graphically adding vectors is the triangle method. This is also called the "head-to-tail" method.

☐ The resultant of two vectors that are 90 degrees apart can be calculated using the Pythagorean Theorem: $c^2 = a^2 + b^2$. ☐ The trigonometric functions of the angles of a right triangle are sine = opposite/hypotenuse; cosine = adjacent/hypotenuse; tangent = opposite/adjacent. ☐ To subtract vectors, the negative of the vectors to be subtracted is added to the other vector. ☐ The negative equivalent of a vector is obtained by rotating the vector 180 degrees.

review questions

1. What is the phase relationship between the voltage and current in a purely inductive circuit?
2. What is the phase relationship between the voltage and current in a purely capacitive circuit?
3. What is the phase relationship between the voltage and current in a purely resistive circuit?
4. How are vectors having the same direction added?
5. How are vectors having opposite directions added?
6. Does the parallelogram method for adding vectors use the head-to-tail method?
7. Is the triangle method for adding vectors a graphical method?
8. What is the *Pythagorean Theorem*?
9. What is meant by the *sine, cosine,* and *tangent* of an angle?
10. How are vectors subtracted?

separating vectors into components

Every vector can be separated, or *resolved,* into two other vectors that are 90 degrees apart, and which when added will produce the original vector. These two vectors are called the *components* of the original vectors. One of them is in the horizontal direction, and is called the *horizontal component;* the other is in the vertical direction, and is called the *vertical component.* In effect, the two components divide the magnitude of the original vector, and show how much of the magnitude is in the horizontal direction and how much is in the vertical direction.

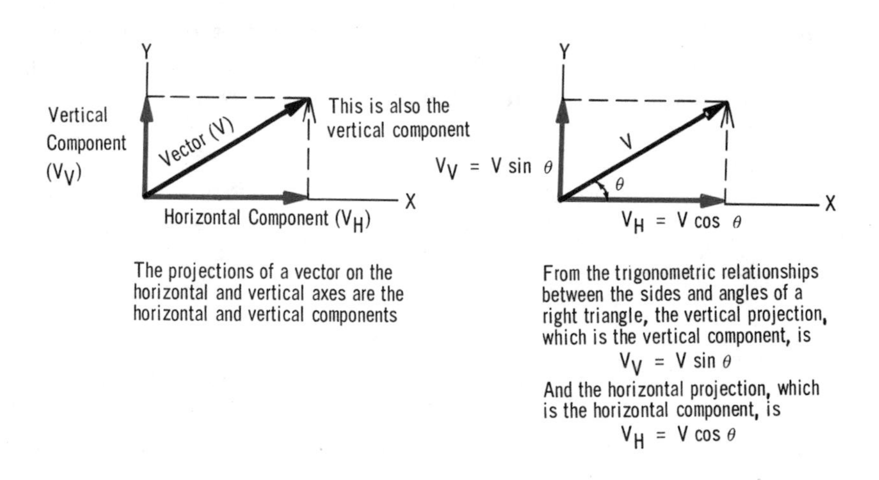

As illustrated, a vector and its two components form a right triangle. You will remember that, in a right triangle, if you know the length of the hypotenuse and one of the angles besides the right angle, you can calculate the lengths of the other two sides using the sine and cosine functions. Therefore, if you know the magnitude and direction of a vector, you can use the sine and cosine functions to find its components. The horizontal component is found by:

$$\text{Horizontal component} = \text{original vector} \times \cos \theta$$

And the vertical component by:

$$\text{Vertical component} = \text{original vector} \times \sin \theta$$

where θ is the angle between the original vector and the horizontal axis.

adding vectors by components

You may wonder why it would ever be desirable to resolve vectors into their components. The primary reason is that it makes the addition of vectors much easier. When vectors are to be added, they can first be resolved into their components. The individual horizontal components can then be added to find the *total horizontal component*. Similarly, the vertical components can be added to find the *total vertical component*. These total components are the components of the resultant vector, and are 90 degrees apart. So the resultant can be found by adding the total components using the standard methods for solving right triangles.

The previous method is not used for adding vectors that are in the same or opposite directions or are 90 degrees apart, inasmuch as relatively simple methods already exist for adding these vectors.

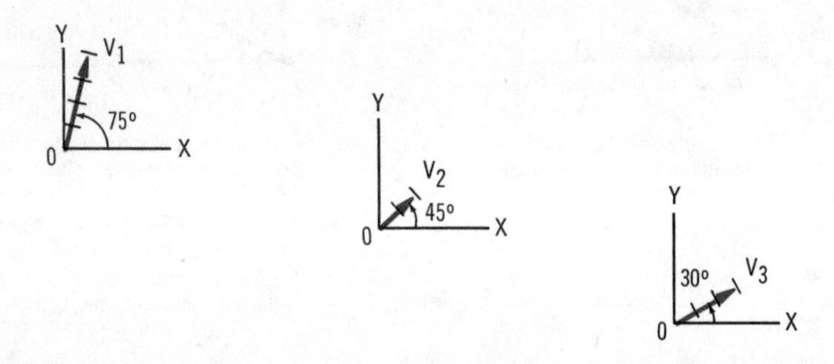

As an example of adding vectors by components, the three vectors shown will be resolved into their components and then added. The horizontal component of each vector is found using the equation $V_H = V \cos\theta$. Therefore,

$$V_{H1} = 4 \times \cos 75° = 4 \times 0.2588 = 1.14$$

$$V_{H2} = 2 \times \cos 45° = 2 \times 0.7071 = 1.41$$

$$V_{H3} = 3 \times \cos 30° = 3 \times 0.866 = 2.6$$

The vertical component of each vector is found next, using the equation $V_V = V \sin\theta$.

$$V_{V1} = 4 \times \sin 75° = 4 \times 0.9659 = 3.86$$

$$V_{V2} = 2 \times \sin 45° = 2 \times 0.7071 = 1.41$$

$$V_{V3} = 3 \times \sin 30° = 3 \times 0.5000 = 1.5$$

adding vectors by components (cont.)

The total components are now found by adding the individual components. The horizontal component, V_H, is

$$V_H = V_{H1} + V_{H2} + V_{H3}$$
$$= 1.14 + 1.41 + 2.6 = 5.15$$

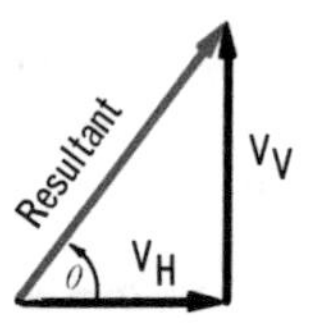

The total vertical component, V_V, is

$$V_V = V_{V1} + V_{V2} + V_{V3}$$
$$= 3.86 + 1.41 + 1.5 = 6.77$$

These total components are now added vectorially to find the magnitude of the resultant vector. Since the components and the resultant form a right triangle, the Pythagorean Theorem is used.

$$c^2 = a^2 + b^2$$
$$c = \sqrt{a^2 + b^2} = \sqrt{(5.15)^2 + (6.77)^2}$$
$$= \sqrt{26.5 + 45.8} = \sqrt{72.3} = 8.5$$

The magnitude of the resultant is thus 8.5 units. Before the vector can be drawn, though, its direction must be determined. This is done by using the relationship between the tangent of the angle enclosed by the resultant and the horizontal axis, and the two components.

Once the tangent of the angle is known, the angle itself can be found from a table of trigonometric functions. In this case, the angle is approximately 53 degrees. The sum of the three original vectors, therefore, is a vector that makes an angle of 53 degrees with the horizontal axis, and that has a magnitude of 8.5 units.

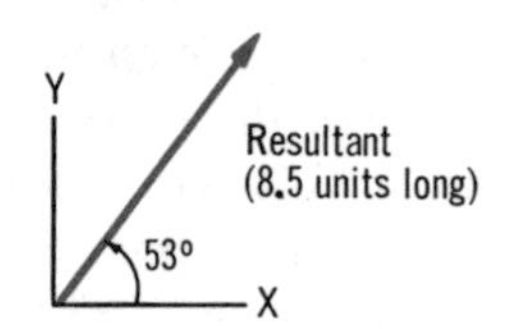

vectors associated with sine-wave phase angles

As was pointed out at the beginning of this volume, vectors are very useful for expressing the phase difference, or phase angle, between a-c voltages and currents. When used in this way, the a-c voltage or current is the physical quantity represented by the vector. As with all vectors, the length of, say, a voltage vector corresponds to the amplitude of the voltage. The direction of the vector, however, does not represent the direction of the voltage. Actually, the direction of the vector

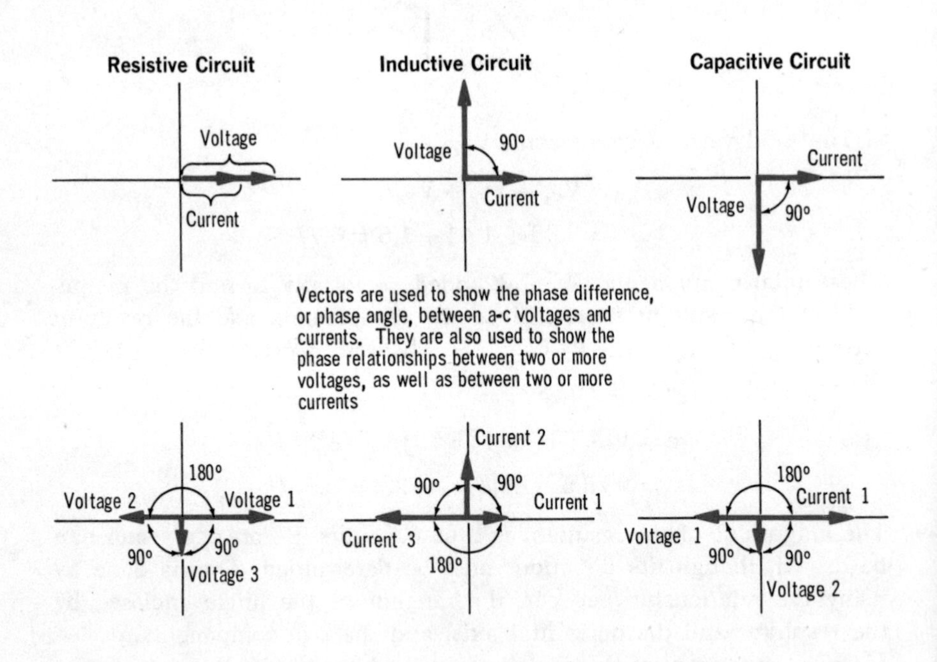

Vectors are used to show the phase difference, or phase angle, between a-c voltages and currents. They are also used to show the phase relationships between two or more voltages, as well as between two or more currents

is *meaningless* in itself, and only becomes significant when it is compared to the direction of another voltage or current vector. Thus, the difference in direction between a-c voltage and current vectors is what is important. This difference is expressed in degrees, and represents the *phase difference* between the two vector quantities.

A-c vectors are called *rotating vectors*, since they represent sine-wave quantities, which as you have learned are based on the rotation of the armature of an a-c generator.

graphical representation of rotating vectors

A-c, or rotating, vectors show both the amplitudes and the phase relationships between sine-wave voltages and currents. The *length* of the vector shows amplitude, while the *angle* between the vectors shows the phase. Also, the position of the vectors shows which is *leading* and which is *lagging*.

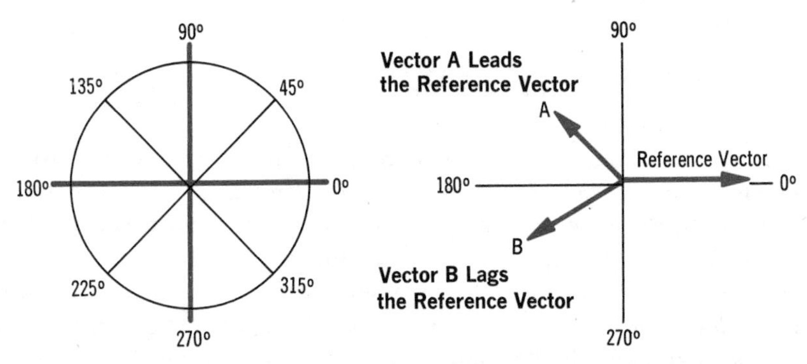

A-c vectors can have a direction corresponding to any angle from 0 to 360° measured from the reference direction of 0°

In any graph of a-c vectors, one vector serves as the reference vector, and points in the 0° direction. Vectors above the horizontal axis lead the reference vector, and those below lag it

As you can see, a vector graph is divided into 360 degrees, corresponding to one full sine-wave cycle. The starting, or *reference*, point of zero degrees is horizontally to the right, and the phase angles of other vectors are compared to this reference. Vectors that *lead* the reference vector are *above* the horizontal axis, and those that *lag* are *below*.

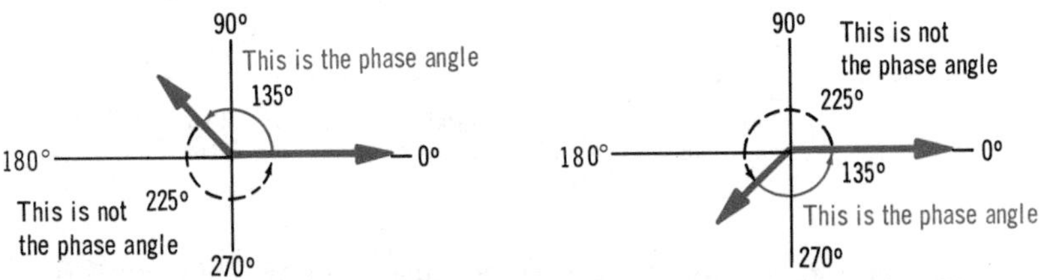

The angle between any vector and the reference vector is measured in the direction that will make it less than 180°. Thus, the phase angle between the reference vector and a vector that leads it is measured in a counterclockwise direction starting from the reference vector. The angle between the reference vector and a vector that lags it is measured in a counterclockwise direction starting from the lagging vector

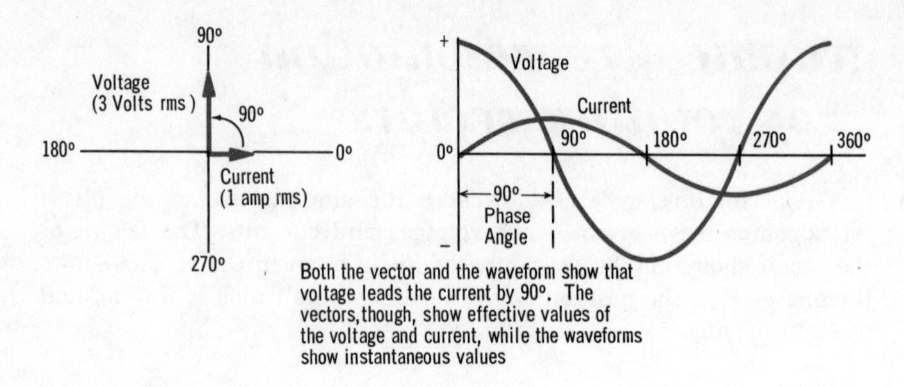

Both the vector and the waveform show that voltage leads the current by 90°. The vectors, though, show effective values of the voltage and current, while the waveforms show instantaneous values

vectors vs. waveforms

Both a-c vectors and waveforms show the amplitudes and phase relationships of a-c voltages and currents. One important difference, though, is in the *type* of amplitudes shown. A waveform shows all of the *instantaneous* values of current or voltage throughout the complete sine-wave cycle. A vector, on the other hand, shows only *one* value, since its length is fixed. This value can be the peak, average, or effective value, depending on the particular situation in which the vector is used. If the length of the vector is proportional to the *peak* amplitude, its *vertical component* is proportional to the instantaneous amplitude for any given phase angle. Therefore, if a vector corresponding to the peak amplitude of a voltage or current is rotated through 360 degrees, as shown below, the magnitude of its vertical component will trace out the waveform that corresponds to the vector.

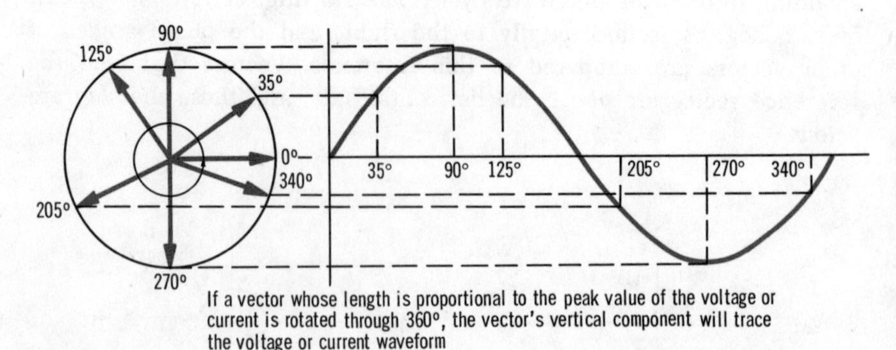

If a vector whose length is proportional to the peak value of the voltage or current is rotated through 360°, the vector's vertical component will trace the voltage or current waveform

You should not think that just because vectors show only a fixed value of voltage or current that their use is thereby limited. They can represent average, effective, or peak amplitudes, and in solving practical problems, these are the values you will usually be working with.

vectors of purely resistive, inductive and capacitive circuits

The following are vector representations, and the corresponding waveforms, of the currents and voltages in purely resistive, inductive, and capacitive circuits.

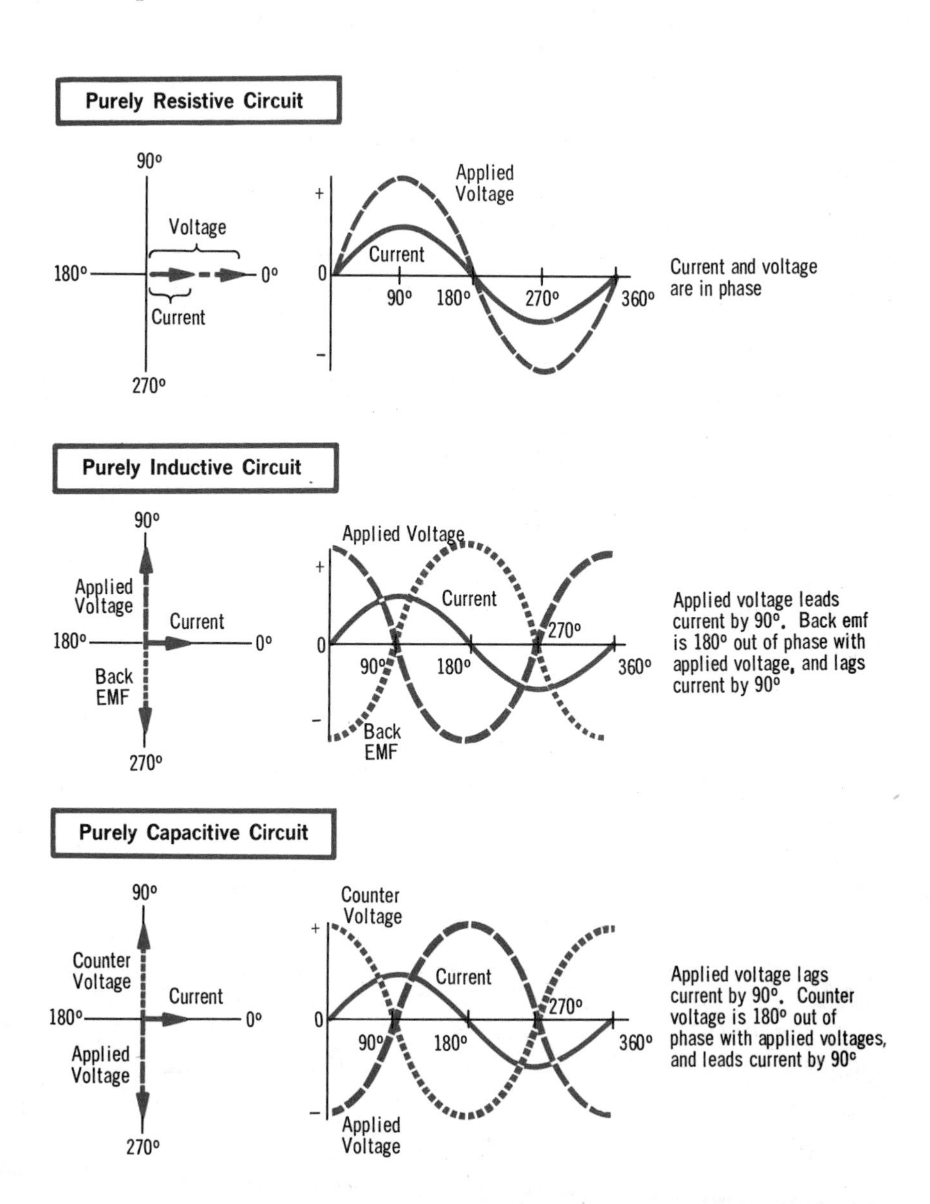

solved problems

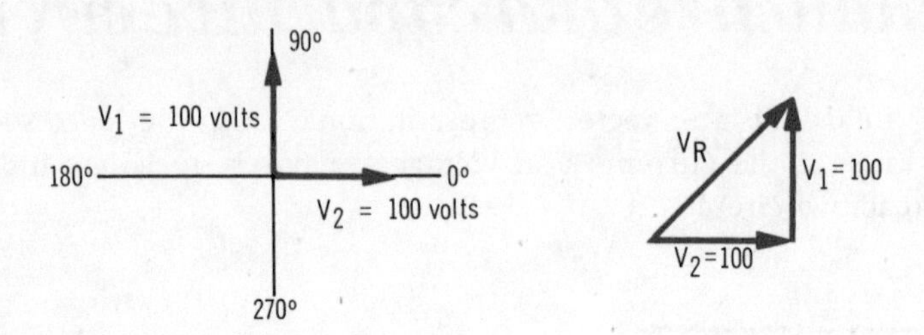

Problem 1. If two voltages with equal amplitudes are 90 degrees apart, as shown, what is their sum?

Since the two voltages are 90 degrees apart, their vectors form a right triangle with the resultant. The Pythagorean Theorem can, therefore, be used to calculate the magnitude of the resultant.

$$V_R^2 = V_1^2 + V_2^2$$
$$V_R = \sqrt{V_1^2 + V_2^2} = \sqrt{(100)^2 + (100)^2} = \sqrt{20{,}000} = 141 \text{ volts}$$

The amplitude of the resultant is therefore 141 volts. Its phase relationship must now be found. This is done using the trigonometric relationship between the tangent of the angle θ, and the two sides of the vector triangle.

V_R = 141
V_1 = 100
V_2 = 100

$$\tan\theta = \frac{\text{opposite side}}{\text{adjacent side}}$$
$$= \frac{V_1}{V_2} = \frac{100}{100} = 1$$

From a table of trigonometric functions, it can be found that the angle whose tangent is equal to 1 is 45 degrees.

The sum of the two original vectors, therefore, is a resultant vector that has an amplitude of 141 volts, and that leads voltage V_2 by 45 degrees.

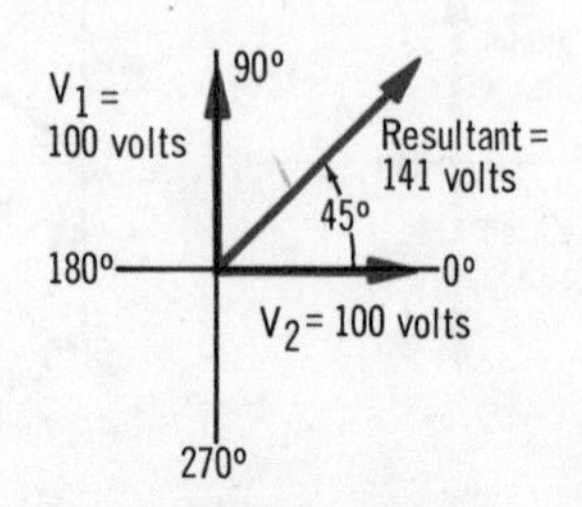

Bookmark

HOWARD W. SAMS & COMPANY

DEAR VALUED CUSTOMER:

Howard W. Sams & Company is dedicated to bringing you timely and authoritative books for your personal and professional library. Our goal is to provide you with excellent technical books written by the most qualified authors. You can assist us in this endeavor by checking the box next to your particular areas of interest.

We appreciate your comments and will use the information to provide you with a more comprehensive selection of titles.

Thank you,

Vice President, Book Publishing
Howard W. Sams & Company

COMPUTER TITLES:

Hardware
☐ Apple 140 ☐ Macintosh 101
☐ Commodore 110
☐ IBM & Compatibles 114

Business Applications
☐ Word Processing J01
☐ Data Base J04
☐ Spreadsheets J02

Operating Systems
☐ MS-DOS K05 ☐ OS/2 K10
☐ CP/M K01 ☐ UNIX K03

Programming Languages
☐ C L03 ☐ Pascal L05
☐ Prolog L12 ☐ Assembly L01
☐ BASIC L02 ☐ HyperTalk L14

Troubleshooting & Repair
☐ Computers S05
☐ Peripherals S10

Other
☐ Communications/Networking M03
☐ AI/Expert Systems T18

ELECTRONICS TITLES:
☐ Amateur Radio T01
☐ Audio T03
☐ Basic Electronics T20
☐ Basic Electricity T21
☐ Electronics Design T12
☐ Electronics Projects T04
☐ Satellites T09
☐ Instrumentation T05
☐ Digital Electronics T11

Troubleshooting & Repair
☐ Audio S11 ☐ Television S04
☐ VCR S01 ☐ Compact Disc S02
☐ Automotive S06
☐ Microwave Oven S03

Other interests or comments: ______________________

Name ______________________
Title ______________________
Company ______________________
Address ______________________
City ______________________
State/Zip ______________________
Daytime Telephone No. ______________________

A Division of Macmillan, Inc.
4300 West 62nd Street
Indianapolis, Indiana 46268

45948

Bookmark

HOWARD W. SAMS & COMPANY

NO POSTAGE
NECESSARY
IF MAILED
IN THE
UNITED STATES

BUSINESS REPLY CARD

FIRST CLASS PERMIT NO. 1076 INDIANAPOLIS, IND.

POSTAGE WILL BE PAID BY ADDRESSEE

HOWARD W. SAMS & CO.
ATTN: Public Relations Department
P.O. BOX 7092
Indianapolis, IN 46209-9921

solved problems (cont.)

Problem 2. Draw the current and voltages in the following circuit on a vector diagram.

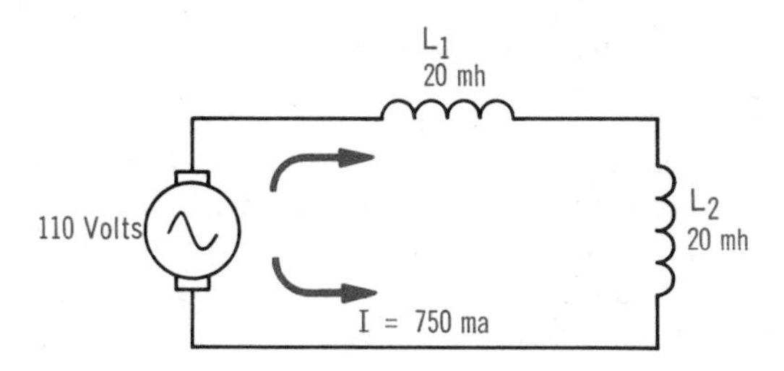

It is a series circuit, so the current throughout the circuit is the same. There will, therefore, be only one current vector, and it should be made the reference vector. The applied voltage will be one voltage vector, and since there are two coils and each develops a back emf, there will be two voltage vectors for the back emf, or a total of three voltage vectors.

Back emf is always 180 degrees out of phase with the applied voltage, so the back emf vectors and the applied voltage vector must have opposite directions. Furthermore, applied voltage in a purely inductive circuit leads the current by 90 degrees. Therefore, if the current is the reference vector and so has a direction of 0 degrees, the applied voltage vector must have a direction of 90 degrees. And since the back emf vectors are opposite in direction to the applied voltage vector, they must have a direction of 270 degrees.

The amplitudes of the applied voltage and the current are stated in the problem; but the amplitudes of the back emf's are not. However, since the inductances of the coils are equal, you do know that the back emf's must also be equal; so their vectors must be equal in length.

Based on the above reasoning, the vector diagram would be drawn as follows:

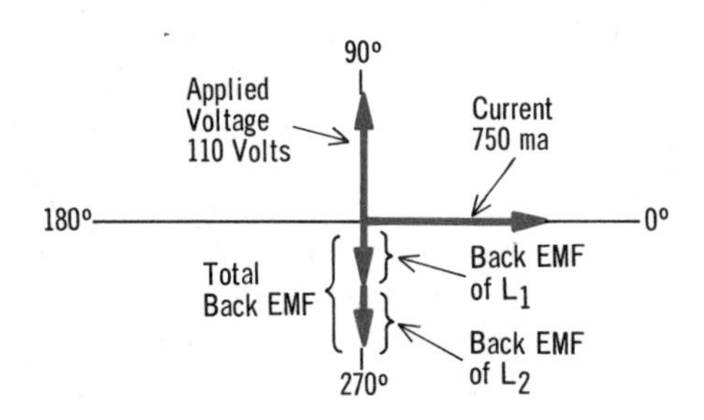

summary

☐ Every vector can be separated, or resolved, into two components that are 90 degrees apart. These are called the horizontal component and the vertical component. ☐ Component vectors in the same direction can be added arithmetically. ☐ The magnitude of the resultant vector can be found from the components by using the Pythagorean Theorem.

☐ A-c vectors are called rotating vectors; they represent sine-wave quantities. ☐ A-c vectors show both the amplitude and the phase relationships between sine-wave voltages and currents. ☐ The length of a rotating vector indicates the amplitude of the a-c voltage or current. ☐ The angle between two rotating vectors indicates the phase difference between the two vector quantities. ☐ A vector graph is divided into 360 degrees, corresponding to one full sine-wave cycle. ☐ Phase angles of vectors are compared to the reference, or zero-degree, point on the right part of the horizontal axis.

☐ A waveform shows all of the instantaneous current and voltage values throughout a complete sine-wave cycle. ☐ A vector shows only one value; it can be the peak, average, or effective value of the amplitude.

review questions

1. Illustrate and explain what is meant by the *horizontal* and *vertical components* of a vector.
2. What is the magnitude of the vertical component of a horizontal vector whose magnitude is 100? What is the magnitude of the horizontal component?
3. For Question 2, what is the phase angle between the vector and the vertical axis?
4. The horizontal component of a vector is 5 and the vertical component is 12. What is the magnitude of the vector? What is its direction with respect to the horizontal axis?
5. The magnitude of a vector is 20 and its horizontal component is 16. What is the value of the vertical component?
6. What is meant by a *rotating vector*? Why is it so called?
7. What does a rotating vector indicate?
8. What amplitude of a voltage or current does a waveform indicate?
9. Explain the difference between a *vector* and a *waveform*.
10. On a vector diagram, show the phase difference between the applied voltage and the current in a purely inductive circuit. In a purely capacitive circuit. In a purely resistive circuit.

RL circuits

An RL circuit is one that contains both *resistance* (R) and *inductance* (L). In Volume 3, you learned the characteristics of a-c circuits having only resistance or only inductance. Many of these characteristics are *modified* when resistance and inductance are *both* present. As a result, different methods and equations must be used to solve RL circuit problems. You will find that the basic reason for the differences between RL circuits and purely resistive or inductive circuits is that the *phase relationships* in the *resistive* portions of RL circuits are *different* than the phase relationships in the *inductive* portions. Both of these relationships, though, affect the overall operation of the circuit, and must be considered when solving RL circuit problems. The nature of the phase relationships that exist in series and parallel RL circuits, as well as methods of solving circuit problems taking into account these phase relationships, are described on the following pages.

Because of their physical construction, every coil contains some resistance, and every resistor contains some inductance. Therefore, all three of these circuits are actually RL circuits

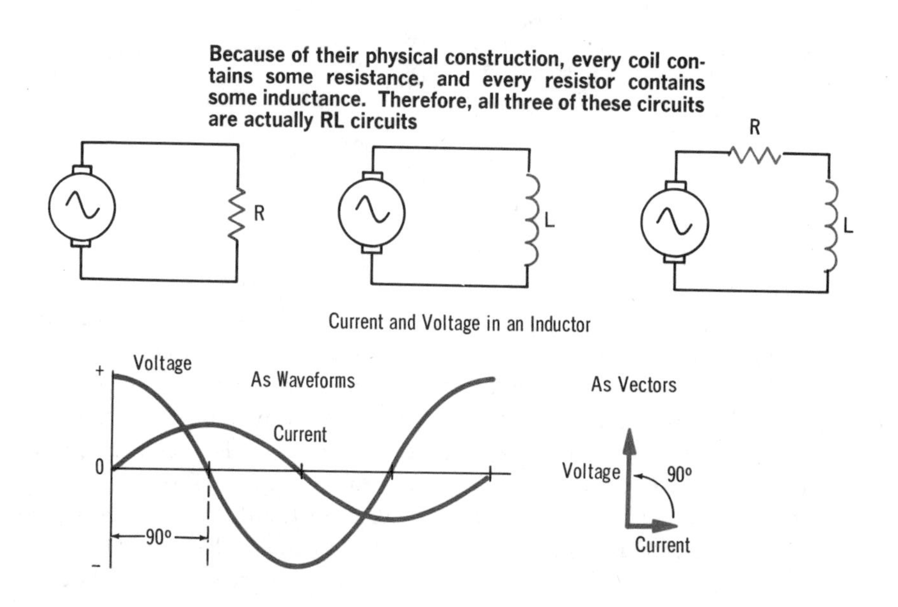

impedance

In resistive circuits, resistance provides the only opposition to current flow. And, in inductive circuits, all of the opposition is provided by the inductance in the form of inductive reactance. Resistance, you will remember, is "built into" a load, and is essentially unaffected by the circuit voltage or current. Inductive reactance, on the other hand, is directly proportional to the frequency, and so its value depends on the frequency of the applied voltage. Furthermore, although a voltage drop occurs when current flows through either a resistance or an inductive reactance, the *phase relationship* between the current and the voltage drop is different for a resistance than it is for a reactance. Since the voltage drop is a measure of the opposition to current, resistance and inductive reactance can be considered as differing in phase. Actually, only quantities that vary in *time* can differ in *phase;* and inasmuch as neither resistance nor inductive reactance do, strictly speaking, they cannot really differ in phase. But, since their *effects* on current flow have a *phase relationship,* we think of resistance and inductive reactance themselves as having a phase relationship.

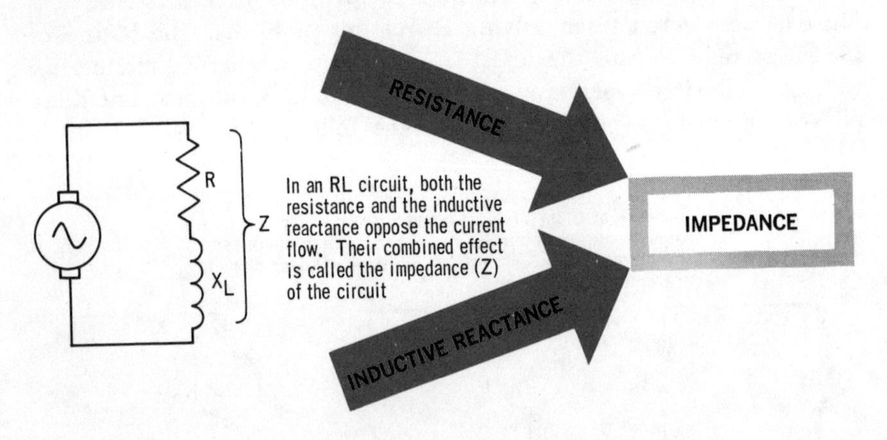

In an RL circuit, both the resistance and the inductive reactance oppose the current flow. Their combined effect is called the impedance (Z) of the circuit

In effect, then, although both resistance and inductive reactance oppose current flow, some of their characteristics and effects are different. For this reason, the *total* opposition to current flow in RL circuits is not expressed in terms of either resistance or inductive reactance. Instead, a quantity called the *impedance* is used. The impedance of an RL circuit is calculated from the values of resistance and inductive reactance, and takes into account the differences between them. Impedance is measured in ohms, and is usually designated by the letter Z. The methods used to calculate impedance depend on whether the resistance and inductive reactance are in *series* or in *parallel.*

series RL circuits

When the resistive and inductive components of a circuit are connected in such a way that the *total* circuit current flows through each, the circuit is a *series* RL circuit. It is important to realize that the current flowing at *every* point in the circuit is the *same*. As you will learn, any analysis of series RL circuits is based on this fact.

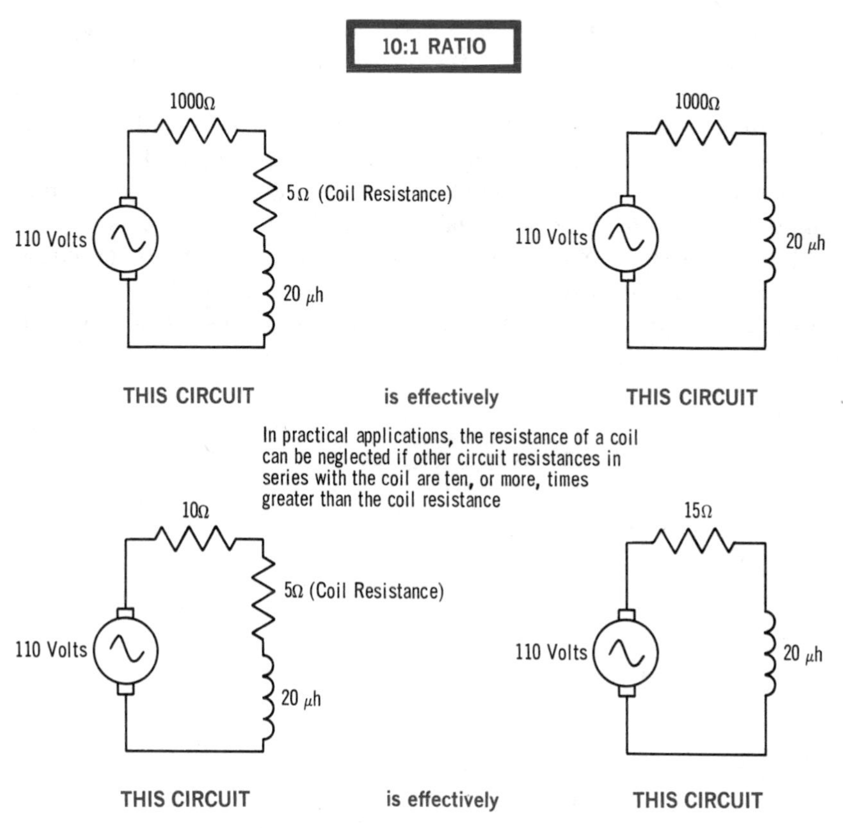

A series RL circuit may consist of one or more resistors, or resistive loads, connected in series with one or more coils. Or, since the wire used in any coil has some resistance, a series RL circuit may consist of only one or more coils, with the resistance of the coils, which is *effectively* in series with the inductance, supplying the circuit resistance.

series RL circuits (cont.)

If a resistor or other resistive load is connected in series with one or more coils, its resistance is usually much greater than the resistance of the coils. When it is *ten,* or more, *times greater,* the effect of the coil resistance can be ignored. In this volume, unless otherwise indicated, we will assume that this is the case, and will consider coils as having zero resistance.

You will notice that in the discussion on series RL circuits, the phase of the circuit *current* is used as the phase *reference* for all other circuit quantities. This is done for convenience, inasmuch as the current is the same throughout the circuit. Since current is used as the reference, the current vector on a vector diagram has a phase angle of *0 degrees,* which means that it is horizontal and points to the right. Any circuit quantity that is in phase with the current, therefore, also has a phase angle of 0 degrees. You should be aware, though, that other circuit quantities could be used as the phase reference. Current is chosen in the series RL circuit only because it is common to all the parts in the circuit.

Use of Current as Phase Reference

I

In series RL circuits, the current is used as the phase reference for all other circuit quantities. It, therefore, has a phase angle of 0°

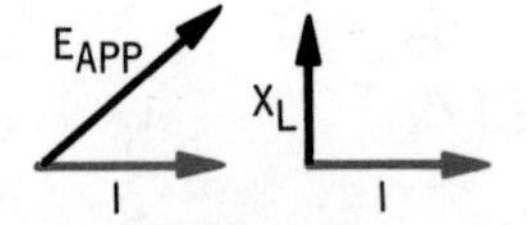

The phase of all other circuit quantities are then determined relative to the current

R and E_R are in phase with I

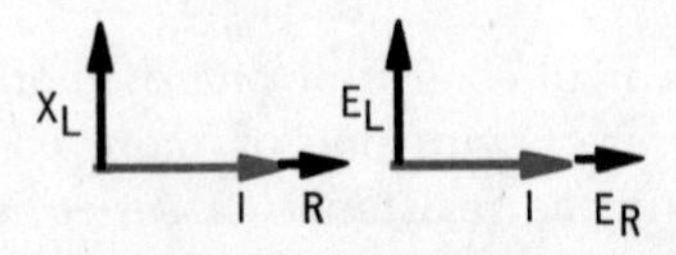

Quantities that are in phase with the current also have phase angles of 0°

voltage

When voltage is applied to a series RL circuit, the current causes a voltage drop across both the resistance and inductance. As you learned in Volume 3, the voltage drop across a *resistance* is *in phase* with the current that causes it, while the voltage drop across an *inductance leads* the current by 90 degrees. Since in a series RL circuit the current through the resistance and inductance is the same, the voltage drop across the resistance (E_R) is in phase with the circuit current, while the voltage drop across the inductance (E_L) leads the current by 90 degrees. With the current as a reference, therefore, *E_L leads E_R* by 90 degrees. The amplitude of the voltage drop across the resistance is proportional to the current and the value of resistance ($E = IR$). The amplitude of the voltage across the inductance is proportional to the current and the value of inductive reactance ($E = IX_L$).

A situation that is new to you, though, arises here. Until now, the *sum* of the individual voltage drops around a series circuit was equal to the applied voltage in all the circuits you have learned. This was according to Kirchhoff's Voltage Law. But if you were to measure the *applied* voltage and then the voltage *drops* in a series a-c circuit, you would find that the *sum* of the voltage drops appears *greater* than the applied voltage. This is because when the voltage drops are not in phase, the *vector sum* of the voltage drops, rather than their arithmetic sum must be used for Kirchhoff's Law to work. Thus, the applied voltage can be represented vectorially as the sum of the two vectors: one, the E_R vector, is at 0 degrees, since it is in phase with the circuit current; the other, the E_L vector, leads both E_R and I by 90 degrees.

The relationship between the applied voltage and the voltage drops in a series RL circuit is such that the applied voltage equals the VECTOR SUM of the voltage drops

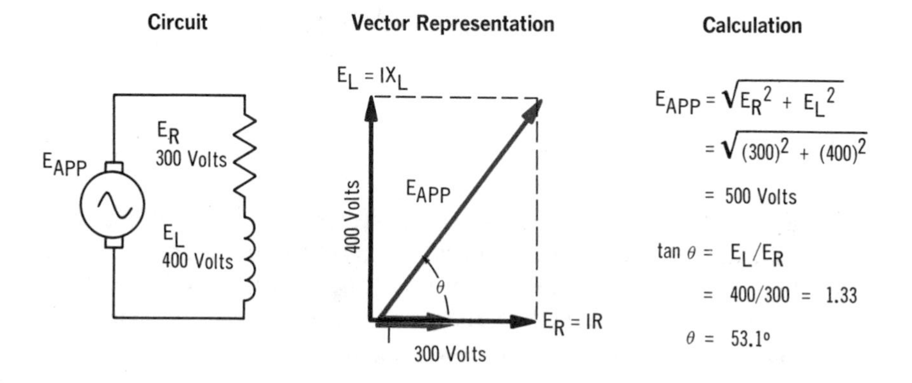

voltage (cont.)

Since the two voltage vectors are 90 degrees apart, they can be added vectorially by the *Pythagorean Theorem* to find the applied voltage:

$$E_{APP} = \sqrt{E_R^2 + E_L^2}$$

Graphically the applied voltage is the hypotenuse of a right triangle whose other two sides are the circuit voltage drops. The angle between the applied voltage and E_R is the same as the phase angle between the applied voltage and the current (I). The reason for this is that E_R and I are in phase. The value of θ can be found from:

$$\tan \theta = E_L/E_R$$

or

$$\cos \theta = E_R/E_{APP}$$

With I used as the reference vector, as shown at the left below, it may seem that the applied voltage changes in phase as the voltage drops change with different values of resistance and inductive reactance. Actually, it only appears this way because the current is used as the 0-degree reference.

I as Reference Vector

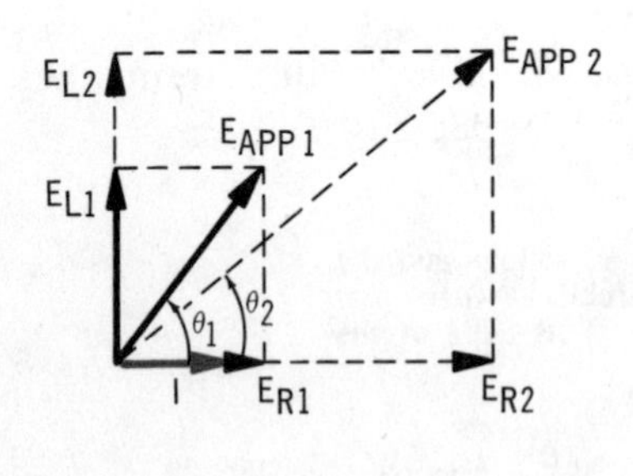

E_{APP} as Reference Vector

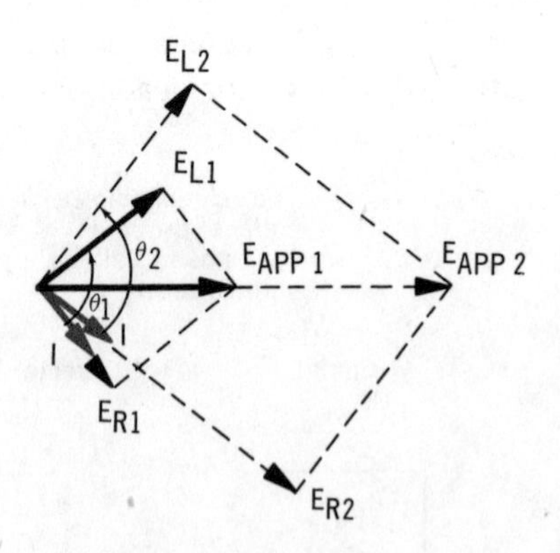

If the applied voltage was used as the reference, you would see that it is the current that actually changes in phase, as shown on the right illustration above. Therefore, to avoid confusion, always consider the phase angle θ as the angle *between* the applied voltage and current, rather than as the phase angle of either one of them.

voltage waveforms

You saw from the calculations on the previous pages that although the individual voltage drops in the circuit were 300 and 400 volts, the applied voltage, or total voltage drop, was 500 volts instead of 700 volts. The reason for this was that the individual voltage drops were out of phase. If they were in phase, they would have reached their maximum amplitudes at the same time, and could be added directly. But since they were out of phase, all of their *instantaneous values* had to be added, and then the *average* or *effective* value of the resulting voltage found. This is what vector addition accomplishes. Whether the average or effective value of the applied voltage is found, depends on what values you are using. If the voltage drops are given in effective values, the applied voltage you find will also be an effective value. Similarly, if you are working with average values, the applied voltage will be an average value.

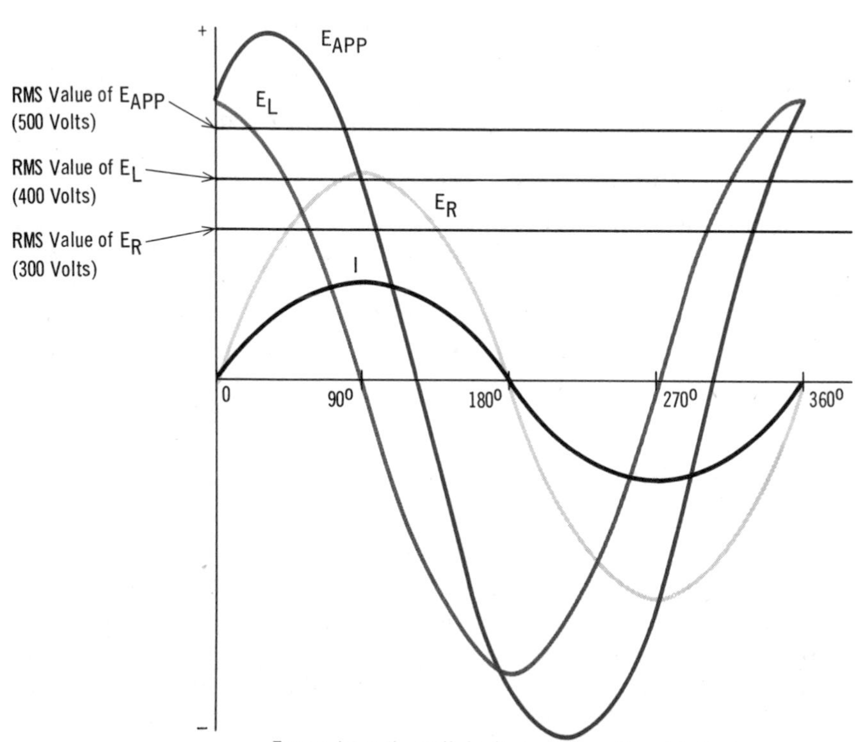

Every point on the applied voltage waveform (E_{APP}) is the algebraic sum of the instantaneous values of the E_R and E_L waveforms. The rms value of the applied voltage waveform is equal to the vector sum of the rms values of the E_R and E_L waveforms

impedance and current

In a series RL circuit, the impedance is the VECTOR SUM of the resistance and the inductive reactance.

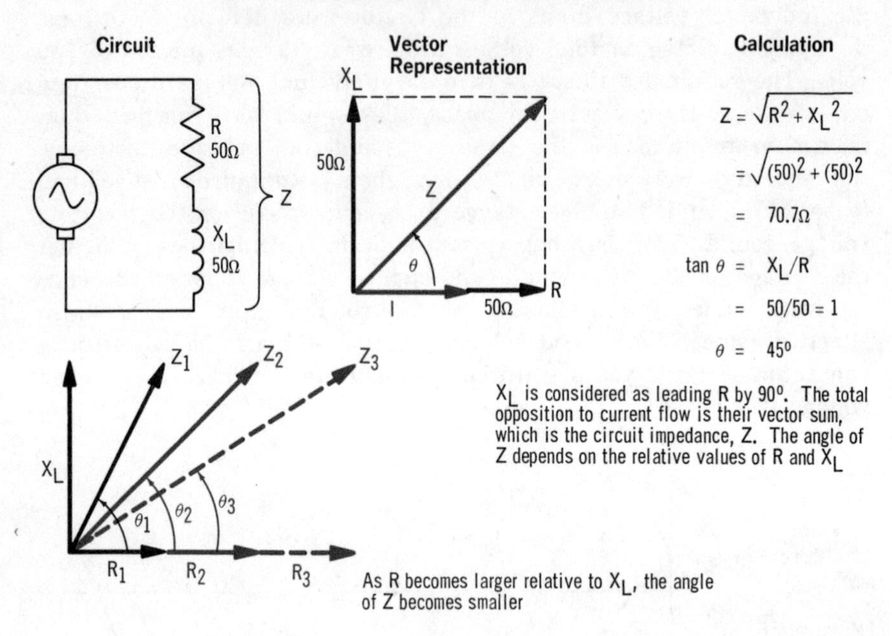

As was explained, it is convenient in RL circuits to consider resistance and inductive reactance as differing in phase, and to use the term impedance (Z) to represent the *total* opposition to current flow. Since, in a series RL circuit, the same current flows through both the resistor and inductor, and the voltage drop across the resistor is in phase with the current, while the voltage drop across the inductor leads the current by 90 degrees, then the inductive reactance is considered to lead the resistance by 90 degrees. The *vector sum* of the reactance and resistance, the impedance, can therefore be calculated by the Pythagorean Theorem.

$$Z = \sqrt{R^2 + X_L{}^2}$$

Since R and X_L are 90 degrees apart, their vector sum, Z, will be at an angle somewhere *between* 0 and 90 degrees relative to the circuit current. The exact angle depends on the *comparative* values of R and X_L. If R is greater, Z will be closer to 0 degrees; and if X_L is greater, Z will be closer to 90 degrees. The angle can be found from:

$$\tan\theta = X_L/R \qquad \text{or} \qquad \cos\theta = R/Z$$

The phase angle of Z is the same as the phase angle of the applied voltage, described previously.

impedance and current (cont.)

The 10-to-1 rule you learned before also applies to impedance. If either X_L or R is 10 times greater than the other, the phase angle of Z can be considered as 0 or 90 degrees, depending on which is larger. Essentially, what the 10-to-1 rule signifies is that if R is 10 or more times greater than X_L, the circuit will operate almost as if X_L were not present; the opposite is true if X_L is 10 or more times larger than R.

The relationships between I, E, and Z in RL circuits are similar to the relationships between I, E, and R in d-c circuits. Because of this, the equations for Ohm's Law can be used in solving RL circuits by using the impedance (Z) in place of the resistance. These equations are often called *Ohm's Law for a-c circuits*. They are:

$$I = E/Z \qquad E = IZ \qquad Z = E/I$$

You will find later that these equations also apply to circuits that contain capacitance as well as inductance and resistance.

In a series RL circuit, the current is the same at every point, and lags the applied voltage by an angle between 0 and 90°

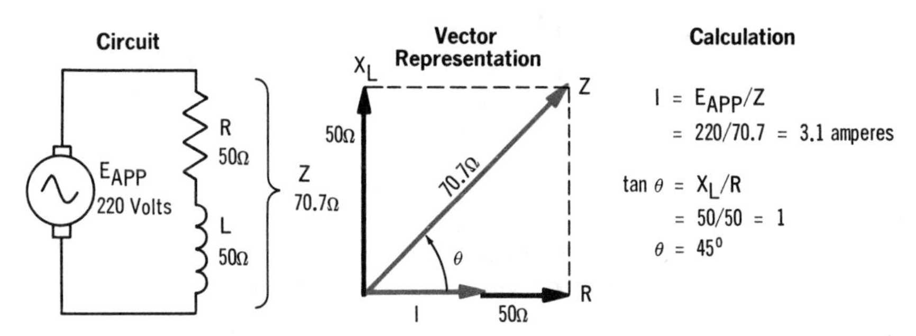

The calculated angle is the phase angle of the current. It is the same as the phase angle of impedance found for the same circuit on the previous page. The reason for this is that the values of X_L and R determine the angle of the impedance, which in turn determines how inductive or resistive the current is

As in any series circuit, the current in an RL series circuit is the *same* at *every point*. As a result, the current through the resistance is *in phase* with the current through the inductance, since they are actually the same current. If the applied voltage and impedance in an RL series circuit are known, the current can be calculated by Ohm's Law: $I = E_{APP}/Z$, where E_{APP} equals the applied voltage and Z equals the vector sum of the resistance and inductive reactance ($\sqrt{R^2 + X_L^2}$).

impedance and current (cont.)

Since the current in a series circuit is common to all parts of the circuit, it is used as the 0-degree reference. And the angle between it and the impedance determines whether the current is more inductive or resistive. As you learned before, the angle of Z is somewhere between 0 and 90 degrees, depending on the relative values of the inductive reactance and resistance.

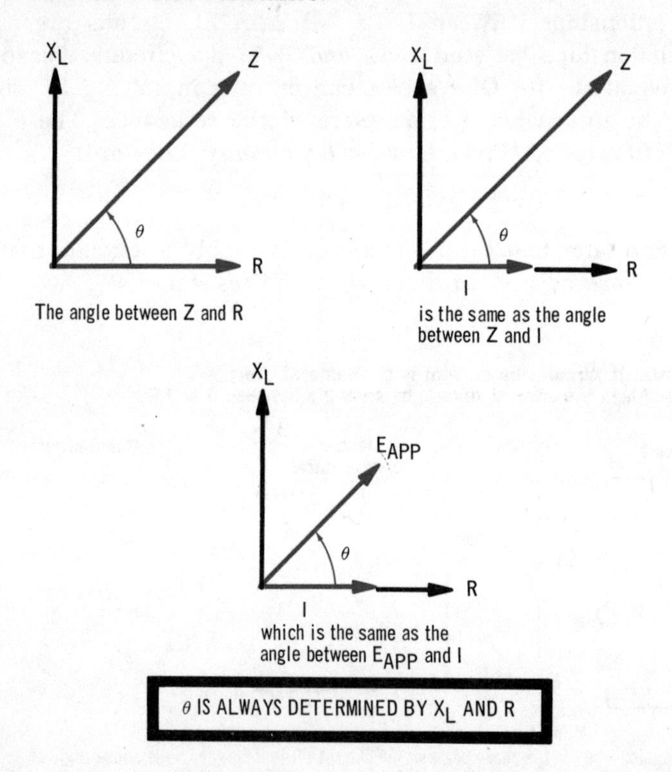

The *greater* the inductive reactance is compared to the resistance, the *larger* is the phase angle, and the more I tends to act as an inductive current. Similarly, the *smaller* the inductive reactance is compared to the resistance, the *smaller* is the angle, and the more I tends to act as a resistive current. When Z and I are exactly in phase ($Z = R$), the current is purely resistive; and when Z leads I by 90 degrees ($Z = X_L$), the current is purely inductive. The terms resistive and inductive, when applied to the current, refer to the phase relationship between the current and the *applied* voltage. The closer the current is to being in phase with the applied voltage, the more resistive it is; and the closer it is to lagging the applied voltage by 90 degrees, the more inductive it is.

power

In resistive circuits, *all* of the power delivered by the source is *dissipated* by the load; but if you recall what you learned from Volume 3, in an RL circuit, only a portion of the input power is dissipated. The part delivered to the inductance is returned to the source each time the magnetic field around the inductance collapses. There are, therefore, two types of power in an RL circuit. One is the *apparent power*. The other type is the *true power* actually consumed in the circuit. The true power depends on how much was returned to the source by the inductance; this in turn depends on the phase angle between the circuit current and the applied voltage. True power is calculated by multiplying the apparent power by the cosine of the phase angle between the voltage and current:

$$P_{TRUE} = E_{APP}I \cos \theta = I^2Z \cos \theta$$

Since from the vector diagram of impedance, $R = Z \cos \theta$, the equation can also be written as $P_{TRUE} = I^2R$. This shows that true power is that used by the circuit resistance.

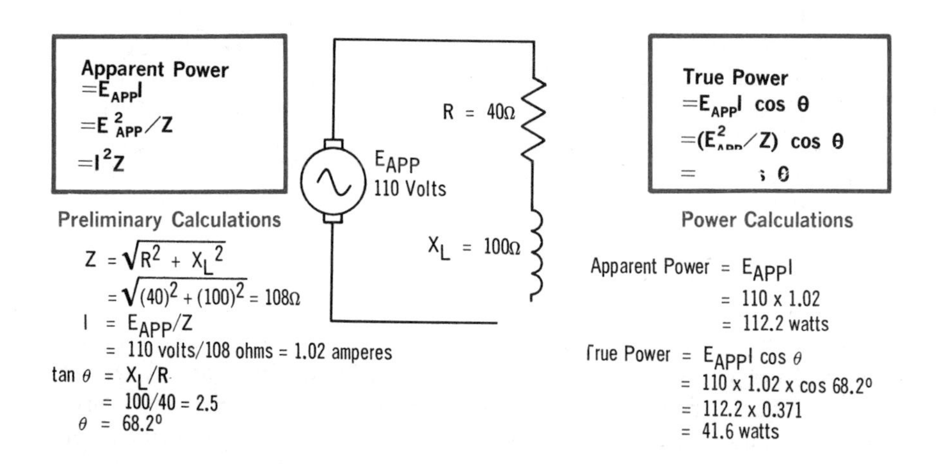

The value of cosine θ can vary between 0 and 1, and, as you remember from Volume 3, it is called the circuit *power factor*. Small power factors (close to 0) are undesirable, since they mean that the power source has to deliver more power than is used. The power factor is found by:

$$\text{Power factor} = \frac{\text{true power}}{\text{apparent power}}$$

the Q of a coil

As you already know, every coil has some resistance, and so acts as a series RL circuit when it is connected to a source of voltage. Physically, it is impossible to measure *separately* the voltage drop across the coil resistance and the drop across the coil inductance. However, mathematically you can assume that the resistance and the inductance are both separate quantities in series with each other, and the two voltage drops and their phase angle can then be calculated. It is obvious that the smaller the coil resistance, the more the coil acts as a *perfect inductor,* that is, one with inductance, but zero resistance. You will find that it is sometimes desirable to compare coils on the basis of how close they approach the theoretically perfect coil. This is done by calculating the *ratio* of inductive reactance to resistance. Such a ratio is called the merit rating, or Q, of the coil. As an equation,

$$Q = X_L/R$$

You can see that the higher the inductive reactance, or the lower the resistance, the greater the Q.

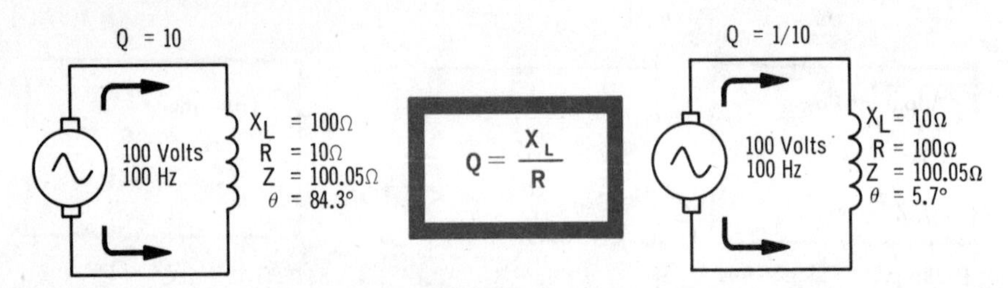

The high-Q coil produces a greater phase angle, and so is a better inductor

A coil with a high Q has a *large phase angle* (close to 90°) between the voltage across it and the current through it. Such a coil will develop a strong magnetic field, and therefore a large cemf, for a given applied voltage. A coil with a low Q has a *small phase angle* between its voltage and current, and because of the I^2R losses caused by its relatively large resistance, will develop a weaker magnetic field and a smaller cemf for a given applied voltage. The Q of a coil becomes an important factor to consider when you work with LC circuits, which are covered later.

With some coils, eddy currents might cause a-c resistance to rise with frequency, as does X_L, to keep the Q somewhat constant. But, since not all coils respond with the same eddy current effects, the Q of different coils should be compared at different frequencies.

effect of frequency

As you know, the *relative values* of X_L and R determine the phase angle of the impedance and the current, as well as the circuit power factor. When X_L is much larger than R, the circuit is very inductive and the power factor is close to zero. When R or X_L is 10 or more times larger than the other, the circuit can be considered purely resistive or purely inductive, and the power factor considered as one or

The characteristics of a series RL circuit vary with different frequencies

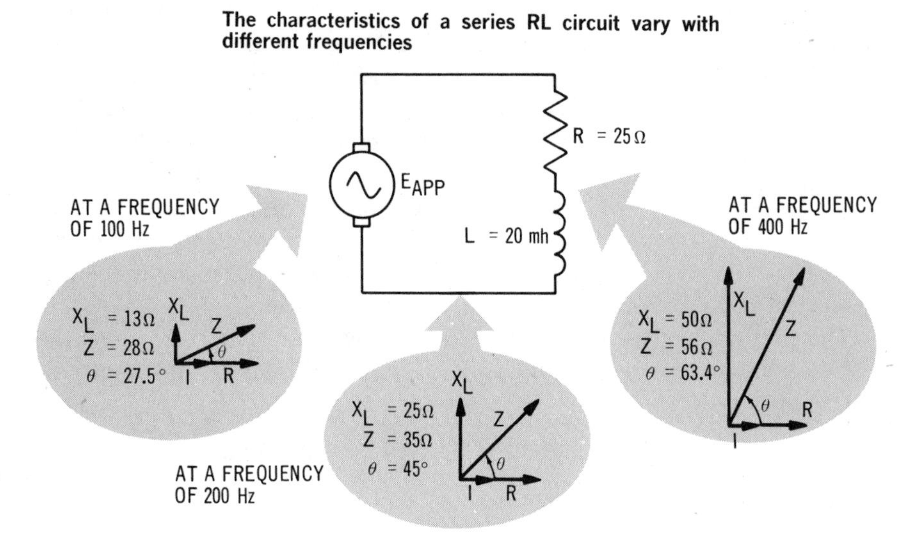

zero. However, since the value of X_L increases with frequency, so might the relative value of X_L and R. As a result, the same circuit might have different properties if the frequency is varied. A very low frequency can make the circuit almost purely resistive; while a very high frequency can cause it to be almost purely inductive. Of course, Z also changes with the relative value of X_L and R.

Very often a graph is made to show circuit impedance changes with frequency. Such a graph is called the frequency response curve of the circuit. The frequency response curve of the circuit above is shown

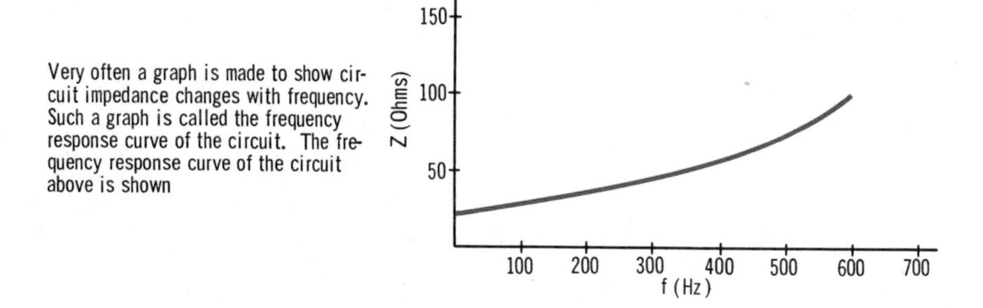

solved problems

Problem 3. What is the current in this circuit?

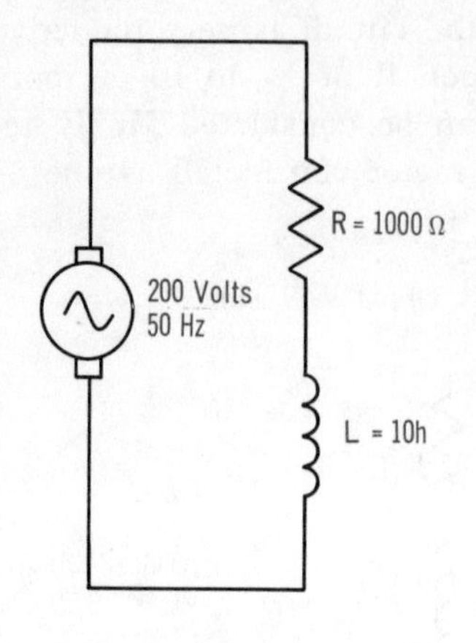

The first step in any problem is to inspect the information to help you determine whether the quantity asked for can be found directly, or if other calculations are required.

The quantity to be solved for is current. The basic equation for current is $I = E/Z$. The voltage is given, so you know that you have to calculate Z. The equation for impedance is $Z = \sqrt{R^2 + X_L^2}$, and from the circuit, you can see that only R is given. Therefore, you also have to solve for the reactance, X_L. The equation for inductive reactance is $X_L = 2\pi fL$, and both f and L are given.

The preliminary inspection, therefore, shows you that three separate calculations are needed to find current: first X_L, then Z, and finally I.

Calculating X_L:

$$X_L = 2\pi fL = 6.28 \times 50 \times 10 = 3140 \text{ ohms}$$

Calculating Z:

$$Z = \sqrt{R^2 + X_L^2} = \sqrt{(1000)^2 + (3140)^2} = 3295 \text{ ohms}$$

Calculating I:

$$I = E/Z = 200 \text{ volts}/3295 \text{ ohms} = 0.061 \text{ ampere}$$

Problem 4. In the above circuit, what is the phase angle between the applied voltage and the current?

The phase angle can be calculated from either of the equations: $\tan\theta = E_L/E_R$ or $\tan\theta = X_L/R$. Since the voltages across the resistance and inductance are not known, but the resistance and inductance are, $\tan\theta = X_L/R$ is used.

$$\tan\theta = \frac{X_L}{R}$$

$$= \frac{3140}{1000}$$

$$= 3.14$$

$$\theta = 72.3°$$

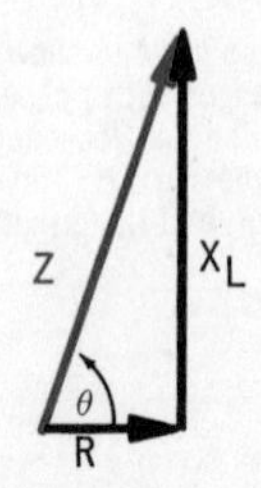

solved problems (cont.)

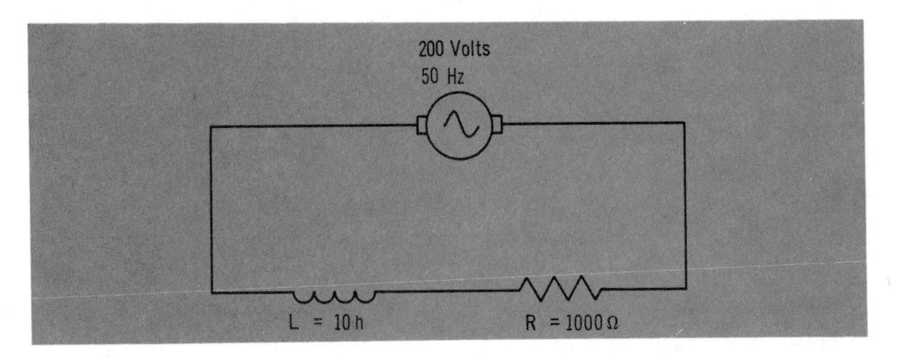

Problem 5. In the circuit, what is the phase angle of the impedance?

The phase angle of the impedance is always identical to the phase angle of the applied voltage and the current. This phase angle was calculated in Problem 4. The angle of the impedance is, therefore, also 72.3°.

Problem 6. In the circuit, what are the voltage drops across R and L?

Both of the voltage drops can be found by Ohm's Law, the drop across the resistance being $E_R = IR$, and the drop across the inductance, $E_L = IX_L$. The values of I, R and X_L are known, so the problem can be solved directly, with no preliminary calculations necessary.

Calculating E_R:

$$E_R = IR$$
$$= 0.061 \text{ ampere} \times 1000 \text{ ohms} = 61 \text{ volts}$$

Calculating E_L:

$$E_L = IX_L$$
$$= 0.061 \text{ ampere} \times 3140 \text{ ohms} = 192 \text{ volts}$$

Problem 7. In the circuit, what would be the voltage drops across R and L if the source frequency were raised to 1000 Hz?

The first step is to find the value that X_L would attain at 1000 Hz by:

$$X_L = 2\pi fL = 6.28 \times 1000 \times 10 = 62{,}800 \text{ ohms}$$

Normally, the next step would be to calculate the impedance and then use E/Z to find the circuit current. Then you would use Ohm's Law to find the voltage drops. But a much simpler method would be to use the 10-to-1 rule for X_L and R, and you can see immediately that the entire source voltage will be dropped across L.

solved problems (cont.)

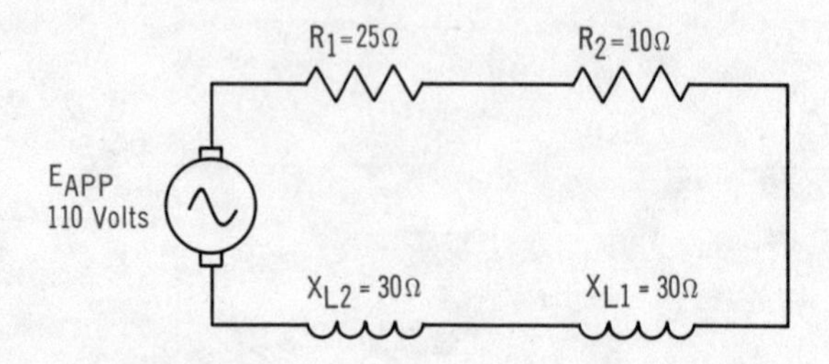

Problem 8. What is the impedance of the circuit?

This circuit is solved in the same manner as one containing only one resistance and one inductance, except that you first must calculate the total resistance and total inductive reactance. Since the resistances and inductances are in series, this is done by simple addition.

$$R_{TOT} = R_1 + R_2$$
$$= 25 + 10 = 35 \text{ ohms}$$
$$X_{L\ TOT} = X_{L1} + X_{L2}$$
$$= 30 + 30 = 60 \text{ ohms}$$

The impedance can then be found by the standard equation:

$$Z = \sqrt{R_{TOT}^2 + X_{L\ TOT}^2}$$
$$= \sqrt{(35)^2 + (60)^2} = \sqrt{4825}$$
$$= 69.5 \text{ ohms}$$

Problem 9. How much power is consumed in the above circuit?

Power consumed, or dissipated, is true power, so the true power must be found. The applied voltage is given and the impedance has just been found. The most appropriate equation to use, therefore, is $P_{TRUE} = (E^2_{APP}/Z) \cos \theta$. Before this equation can be applied, though, the phase angle, θ, must be found.

$$\tan \theta = \frac{X_{L\ TOT}}{R_{TOT}} = 60/35 = 1.71$$
$$\theta = 59.7°$$

The true power can now be calculated.

$$P_{TRUE} = \frac{E_{APP}^2}{Z} \cos \theta$$
$$= \frac{(110)^2}{69.5} \cos 59.7°$$
$$= 174.1 \times 0.505 = 87.9 \text{ watts}$$

summary

☐ The total opposition to current flow in an RL circuit is called the impedance. Impedance is measured in ohms, and is designated by Z. ☐ In a series RL circuit, the current through each component is the same. Because of this, the current is used as the phase reference for all other circuit quantities. ☐ Kirchhoff's voltage equation is valid for a series RL circuit, but the voltages across the resistor and inductor must be added vectorially.

☐ The applied voltage in a series RL circuit is found by $E_{APP} = \sqrt{E_R^2 + E_L^2}$. ☐ The value of the phase angle (θ) is found by: $\tan \theta = E_L/E_R$; or $\cos \theta = E_R/E_{APP}$. ☐ The magnitude of the impedance is the vector sum of the reactance and the resistance: $Z = \sqrt{R^2 + X_L^2}$. ☐ The phase angle between R and X_L is found by: $\tan \theta = X_L/R$; or $\cos \theta = R/Z$. ☐ Either X_L or R can be considered negligible if one is 10 times as large as the other.

☐ Ohm's Law for a-c circuits can be expressed as: $E = IZ$. Other forms of the equation are $I = E/Z$ and $Z = E/I$. ☐ The actual power dissipated in a circuit is called true power. It is equal to the apparent power times the power factor: $P_{TRUE} = E_{APP}I \cos \theta = I^2Z \cos \theta$. ☐ The Q of a coil is the ratio of the reactance at a particular frequency to the resistance of the coil.

review questions

1. In a vector diagram of the voltages in a series RL circuit, what circuit quantity is used as the reference vector? Why?
2. What is the 10-to-1 rule for R and X_L in a series RL circuit?
3. What is the resistance of a coil having a Q of 65, when the inductive reactance is 325 ohms?
4. What is the resistance in a series RL circuit when the impedance is 130 ohms and the inductive reactance is 50 ohms?
5. What is the phase angle for the circuit in Question 4?
6. If an applied voltage of 100 volts causes 5 amperes of current in a series RL circuit, what is the circuit impedance?
7. The power dissipated in a circuit is 500 watts, the impedance is 10 ohms, and the phase angle is 60 degrees. What is the value of the current in the circuit?
8. What is the apparent power of a circuit which dissipates 500 watts, and has a power factor of 0.25?
9. Between what values can the power factor be found? Why?
10. What is the voltage across the coil in a series RL circuit when the applied voltage is 100 volts, and the voltage across the resistor is 80 volts?

parallel RL circuits

In a parallel RL circuit, the resistance and inductance are connected in parallel across a voltage source. Such a circuit thus has a *resistive branch* and an *inductive branch*. The circuit current *divides* before entering the branches, and a portion flows through the resistive branch, while the rest flows through the inductive branch. The currents in the branches are therefore *different*. The analysis of parallel RL circuits

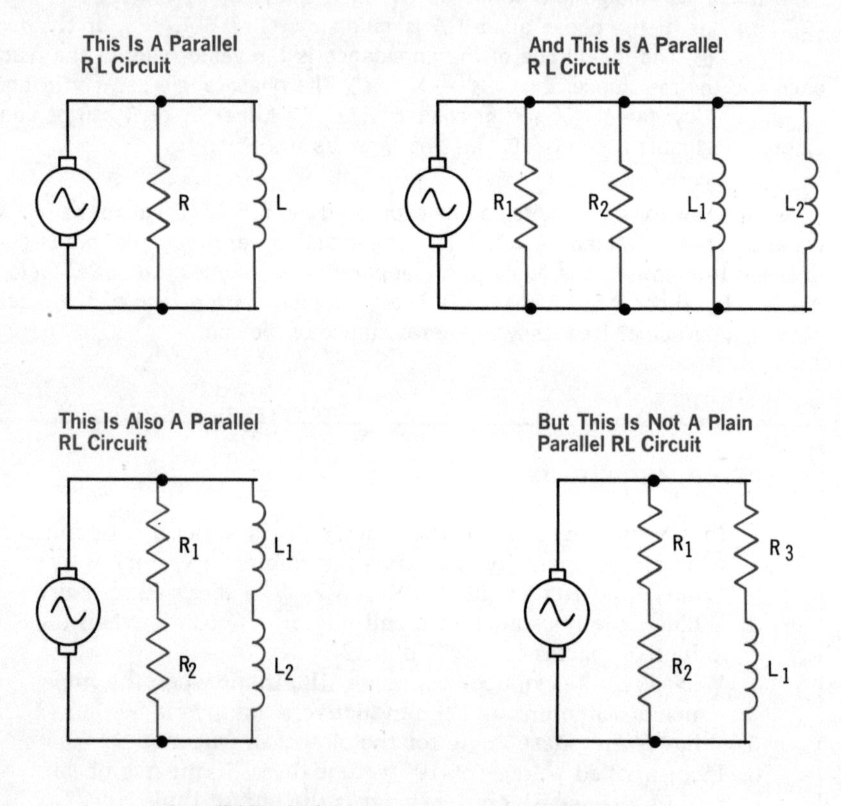

A parallel RL circuit has one or more resistive branches and one or more inductive branches. Each resistive branch is purely resistive, and each inductive branch is purely inductive. If any branch contains both resistance and inductance, the circuit is a series-parallel RL circuit. These will be covered later

and the methods used to solve them are different than the analysis and solution of series RL circuits. It is important, therefore, that you be able to recognize series and parallel RL circuits so that you can use the proper techniques and methods for solving them.

voltage

In a simple parallel RL circuit, there is one *resistive branch* and one *inductive branch.* Both branches are connected directly across the voltage source, and so have the full source voltage applied to them. Since the source voltage across both branches is the *same,* the voltages must be *in phase.* You can see, therefore, that if you know the applied voltage, you automatically know the voltage across each branch. Likewise, if you know the voltage across either branch, you also know the voltage across the other branch, as well as the applied voltage.

The voltage across every branch of a parallel RL circuit is the same as the applied voltage

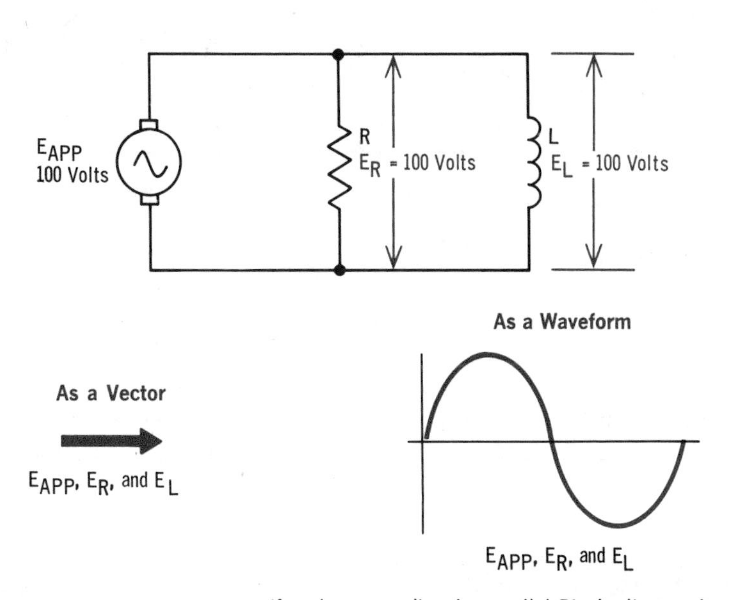

If you know one voltage in a parallel RL circuit, you automatically know the other voltages, since they are all identical

You will remember that in *series* RL circuits, *current* was the common quantity, since it was the same in both the resistive and inductive parts of the circuit. In *parallel* RL circuits, *voltage* is the common quantity, inasmuch as the same voltage is applied across the resistive and inductive branches. The currents in the branches are not the same. The voltage, therefore, is used as the zero reference to compare the other angles.

branch current

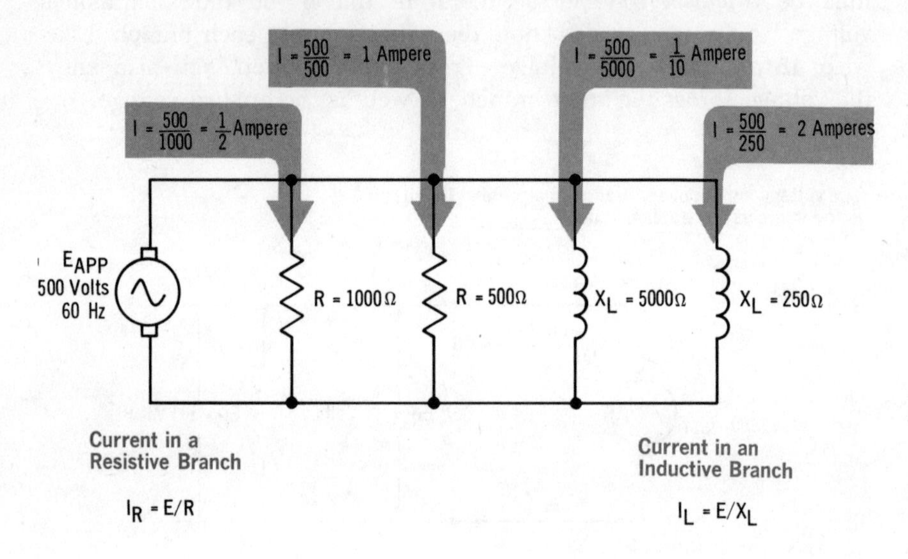

As in all parallel circuits, the current in each branch of a parallel RL circuit is *independent* of the currents in the other branches. If one of the branches opened, it would have no effect on the other branch currents. The current in each branch depends only on the voltage across that branch and the opposition to current flow, in the form of either resistance or inductive reactance, contained in the branch. The branch voltages are the same, so it is the value of resistance or inductive reactance that determines the relative amount of current that flows in the branch. Each branch of a parallel RL circuit can be considered as a small separate series circuit. Ohm's Law can then be used to find the individual branch currents. In resistive branches, therefore, the current is equal to the branch voltage, which is the same as the applied voltage, divided by the resistance. In inductive branches, the current equals the branch voltage divided by the inductive reactance. Thus,

$$I_R = \frac{E}{R} \qquad I_L = \frac{E}{X_L}$$

Line current in a parallel RL circuit is equal to the VECTOR SUM of the currents in the resistive and inductive branches

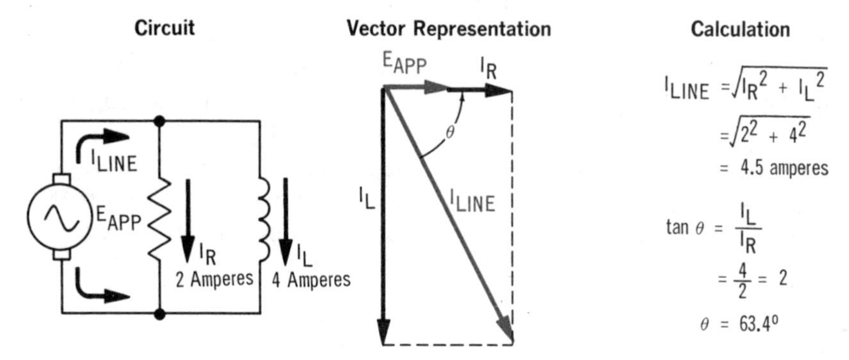

line current

In purely resistive parallel circuits, the total circuit current, or line current as it is called, is simply the *arithmetic sum* of the individual branch currents. In parallel RL circuits, though, there is a *phase difference* between the current in the resistive branch and the current in the inductive branch. Because of the phase difference, the individual branch currents must be added *vectorially* to find the line current. The nature of the phase difference between the two currents is such that the current in the *resistive* branch *leads* the current in the *inductive* branch by 90 degrees. The reason for this is that the voltages across the branches are in phase, and the current in the resistive branch is in phase with the voltage, while the current in the inductive branch lags the voltage by 90 degrees.

Because the two currents are 90 degrees out of phase, their vector sum, which is the line current, can be calculated by the Pythagorean Theorem, using the equation:

$$I_{LINE} = \sqrt{I_R^2 + I_L^2}$$

Line current is the vector sum of the resistive and inductive currents

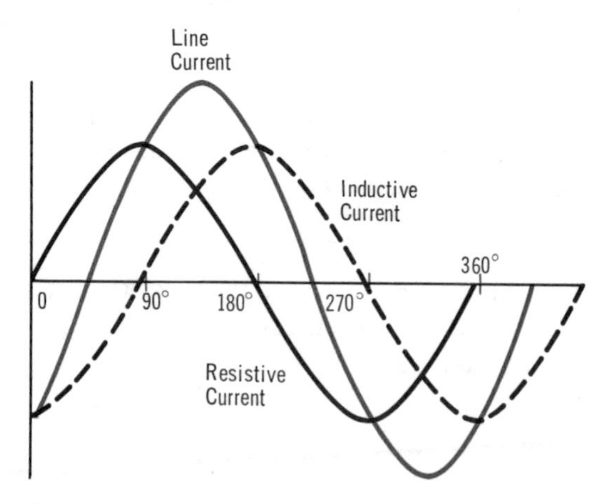

Resistive current leads the inductive current by 90°

line current (cont.)

The phase angle between the line current and the applied voltage is somewhere between 0 and 90 degrees, with the current lagging the voltage, as in all RL circuits. The actual angle depends on whether there is more inductive current or resistive current. If there is more inductive current, the phase of the line current will be closer to 90 degrees. It will be closer to 0 degrees if there is more resistive current.

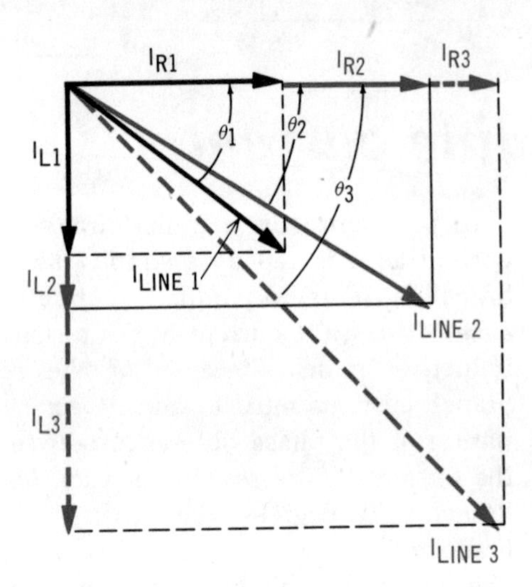

The line current in a parallel RL circuit will have a phase angle between 0 and 90°, lagging. The value depends on the relative values of the inductive and resistive currents in the branches

If either the resistive or inductive current is 10 times or more greater, the line current can be considered to have a phase angle of 0 or 90 degrees, as the case may be. From the vector diagram, you can see that the value of the phase angle can be calculated from the equation:

$$\tan \theta = I_L/I_R$$

Other very useful equations for calculating the phase angle can be derived by substituting the relationships $I_L = E/X_L$ and $I_R = E/R$ into the above equation. The equations derived in this way are

$$\tan \theta = R/X_L \qquad \text{and} \qquad \cos \theta = Z/R$$

If the impedance of a parallel RL circuit and the applied voltage are known, the line current can also be calculated using Ohm's Law for a-c circuits:

$$I_{LINE} = E/Z$$

current waveforms

Since the branch currents in a parallel RL circuit are out of phase, their vector sum rather than their straight arithmetic sum equals the line current. This is the same type of situation that exists for the voltage drops in a series RL circuit. By adding the currents vectorially, you are adding all of their instantaneous values, and then finding the average or effective value of the resulting current. This can be seen from the current waveforms shown. They are the waveforms for the circuit solved on the previous pages.

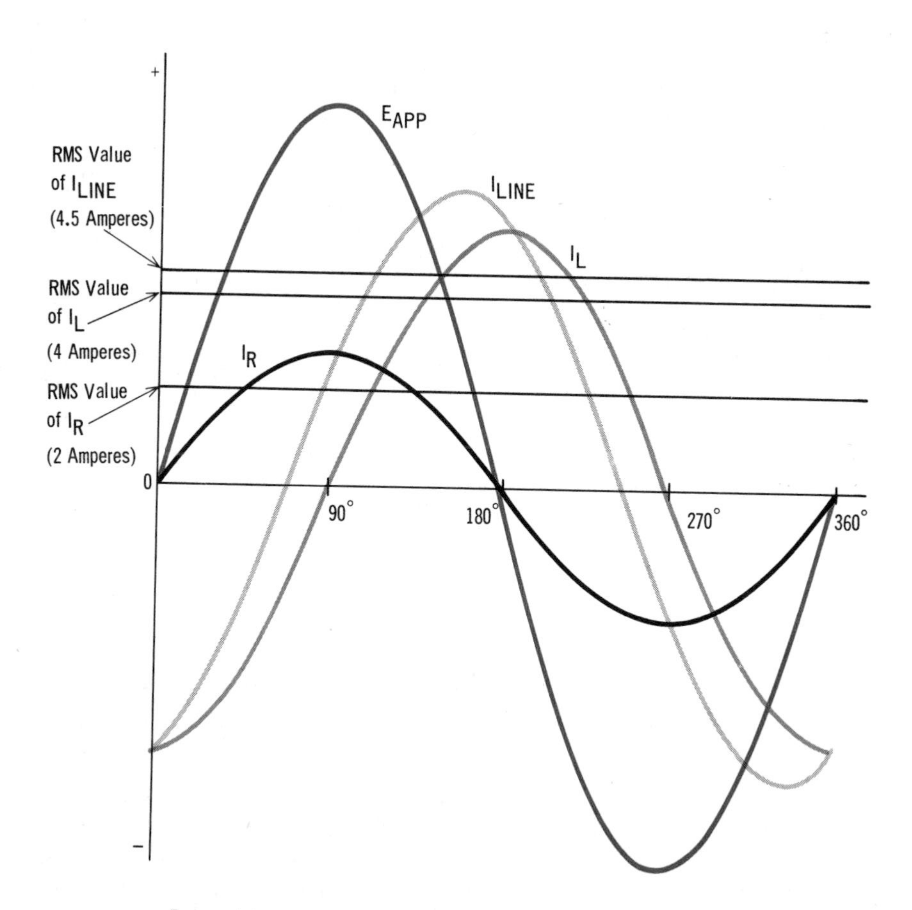

Every point on the line current waveform (I_{LINE}) is the algebraic sum of the instantaneous values of the I_R and I_L waveforms. The rms value of the line current waveform is equal to the vector sum of the rms values of the I_R and I_L waveforms

impedance

The impedance, Z, of a parallel RL circuit is the total opposition to current flow offered by the resistance of the resistive branch and the inductive reactance of the inductive branch. The impedance of a parallel RL circuit is calculated similarly to a parallel resistive circuit. However, since X_L and R are vector quantities, they must be added *vectorially*. As a result, the equation for the impedance of a parallel RL circuit is

$$Z = \frac{RX_L}{\sqrt{R^2 + X_L^2}}$$

where the quantity in the denominator is the vector addition of the resistance and the inductive reactance. If there are more than one resistive or inductive branches, R and X_L must equal the *total* resistance or reactance of these parallel branches.

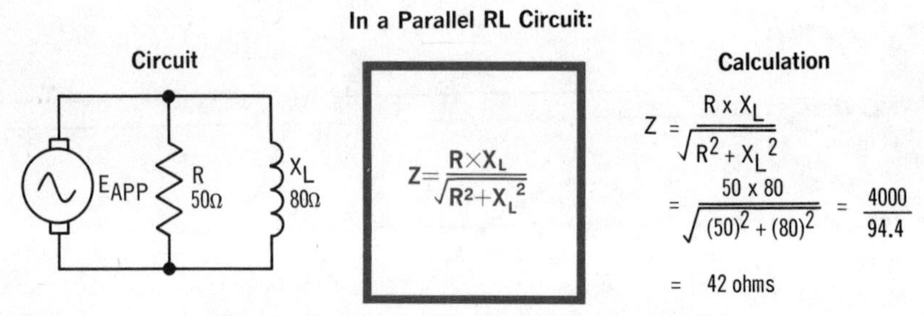

The impedance of a parallel RL circuit is always less than the resistance or inductive reactance of any of the branches

If the circuit line current and the applied voltage are known, the impedance can also be calculated by the equation:

$$Z = E_{APP}/I_{LINE}$$

The impedance of a parallel RL circuit is always *less* than the resistance or reactance of any one branch. The branch of a parallel RL circuit that offers the *most opposition* to current flow has the *lesser effect* on the phase angle of the current. For example, if X_L is larger than R, the resistive branch current is greater than the inductive branch current, so the line current is also more resistive (closer to 0°). This is the opposite of a series RL circuit. When either X_L or R is 10 or more times greater than the other, for practical purposes, a parallel RL circuit can be considered as a simple series circuit consisting of only X_L or R, whichever is *smaller*.

power

In parallel RL circuits, the relationships between the applied voltage, the circuit current, and the circuit power are similar to those of series RL circuits, previously described. Because of the phase difference between the branch currents, the line current and the applied voltage are out of phase. As a result, the value of power obtained by multiplying the applied voltage by the line current is only the *apparent power.* Part of this apparent power is *returned* to the source by the inductive branch. So, to find the power actually dissipated in the circuit, the *true power,* the apparent power has to be multiplied by the cosine of the phase angle (θ) between the applied voltage and line current. The value of cosine θ is the circuit *power factor.*

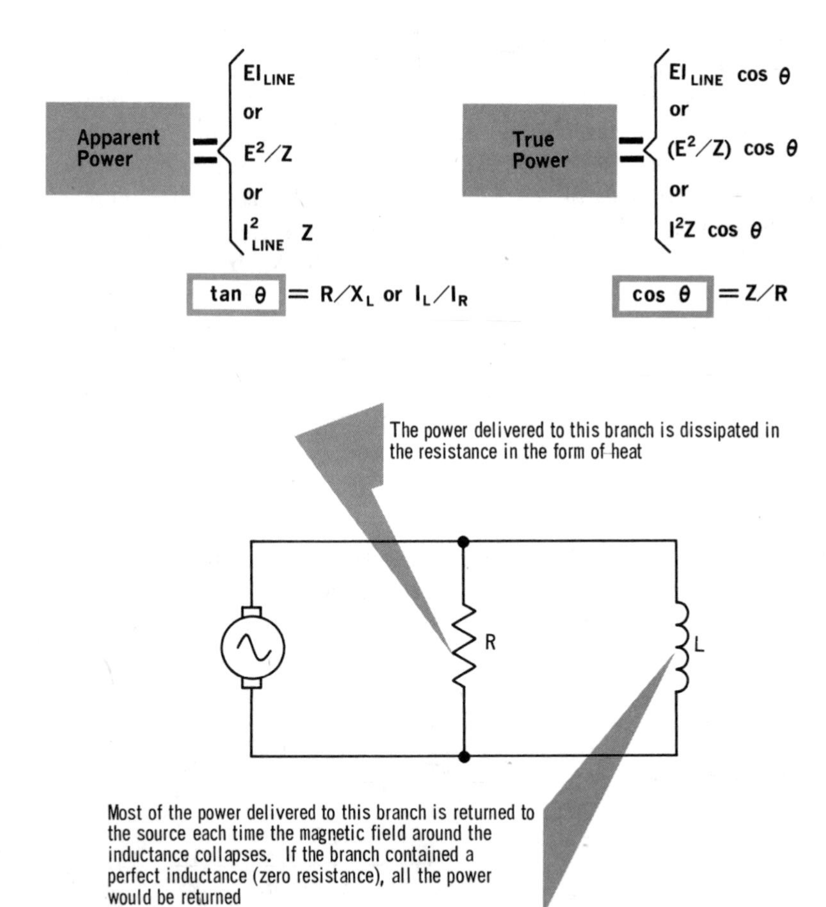

The characteristics of a parallel RL circuit vary with different frequencies

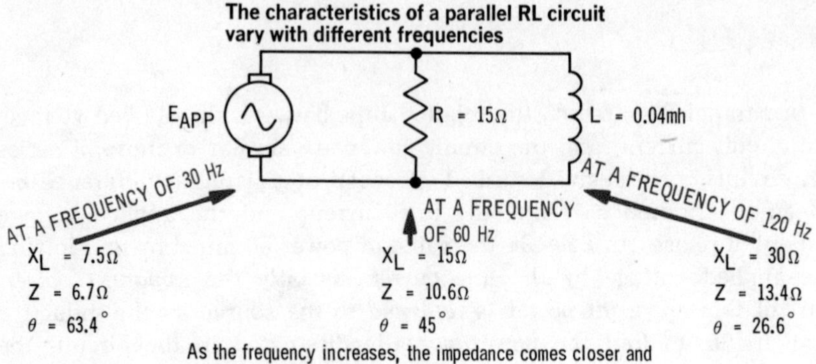

As the frequency increases, the impedance comes closer and closer to the value of the resistance

effect of frequency

You will recall that the frequency of the applied voltage has a significant effect on the characteristics of a series RL circuit. The same thing is true of parallel RL circuits, but the *effects* of frequency changes are *different*. In a series circuit, an increase in frequency caused an increase in the values of X_L and Z, and made the circuit more inductive. Increasing the frequency of a parallel RL circuit also causes an increase in the values of X_L and Z. However, whereas in a *series* circuit a larger X_L makes the circuit *more inductive*, increasing X_L in a *parallel* circuit makes the circuit *more resistive*. The reason for this is that the larger X_L is, the smaller is the inductive branch current, and so the larger is the relative value of the resistive branch current.

If the frequency is decreased, the opposite is true. X_L becomes smaller, causing an increase in inductive branch current, thereby making the circuit more inductive. At very *low frequencies*, therefore, a parallel RL circuit will be almost purely *inductive;* while at very *high frequencies*, it will be almost purely *resistive*. If the frequency is such that either X_L or R are 10 or more times larger than the other, the branch containing the larger one can be neglected, and the circuit treated as a series circuit containing only the smaller of the two.

The frequency response curve of the above circuit

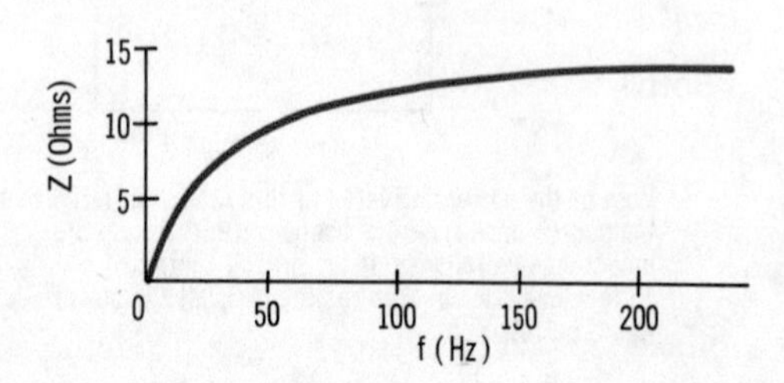

solved problems

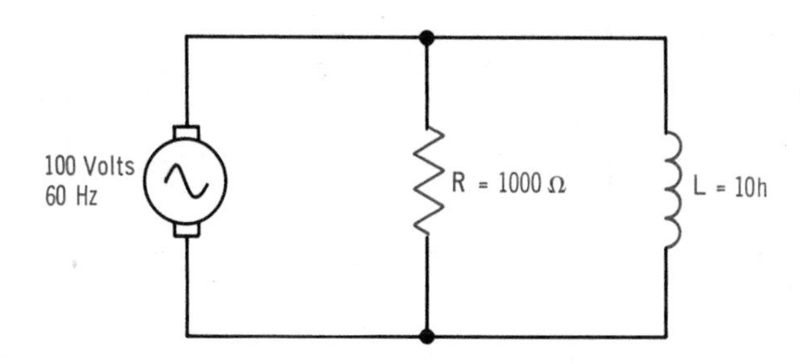

Problem 10. ***Calculate the line current in this circuit in two ways: first using Ohm's Law, and then using the branch currents.***

To calculate the line current using Ohm's Law, the equation $I_{LINE} = E/Z$ is used. Before this can be done, though, the impedance must be found. The equation for the impedance is $Z = RX_L/\sqrt{R^2 + X_L^2}$. R is given in the problem, but X_L is not. So X_L must be found first. This means that to solve the problem, X_L must first be calculated, then Z, and finally the line current.

Calculating X_L:

$$X_L = 2\pi fL = 6.28 \times 60 \times 10 = 3768 \text{ ohms}$$

Calculating Z:

$$Z = \frac{RX_L}{\sqrt{R^2 + X_L^2}} = \frac{1000 \times 3768}{\sqrt{(1000)^2 + (3768)^2}} = 966 \text{ ohms}$$

Calculating I_{LINE}:

$$I_{LINE} = E/Z = 100 \text{ volts}/966 \text{ ohms} = 0.104 \text{ ampere}$$

To calculate the line current using the branch currents, you must first find the two branch currents. The vector sum of these two currents will then be the line current.

Calculating Branch Currents:

$$I_R = E/R = 100 \text{ volts}/1000 \text{ ohms} = 0.1 \text{ ampere}$$

$$I_L = E/X_L = 100 \text{ volts}/3768 \text{ ohms} = 0.0265 \text{ ampere}$$

Calculating I_{LINE}:

$$I_{LINE} = \sqrt{I_R^2 + I_L^2} = \sqrt{(0.1)^2 + (0.0265)^2} = 0.104 \text{ ampere}$$

Both methods, therefore, result in the same value for the line current, which they naturally should. You will find that many, if not most, electrical problems can be solved more than one way. When solving such problems, it is good practice, if time permits, to solve the problems in two ways. The two answers should be the same, and will serve as a check on the accuracy of your calculations.

solved problems (cont.)

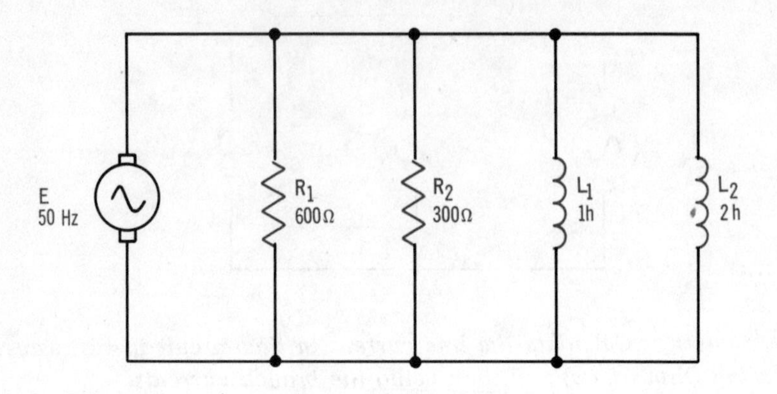

Problem 11. In this circuit, what is the phase angle between the applied voltage and the line current?

The phase angle can be calculated from the equation $\tan\theta = I_L/I_R$, $\tan\theta = R/X_L$, or $\cos\theta = Z/R$. Since the applied voltage is not known, the equation $\tan\theta = I_L/I_R$ cannot be used in this problem. Of the other two equations, $\tan\theta = R/X_L$ is the easiest to use in this particular case. However, before it can be used to solve for the phase angle, the total values of R and X_L must be determined.

Calculating R_{TOT}: The two resistances are in parallel, so the product/sum method is used to find their total.

$$R_{TOT} = \frac{R_1R_2}{R_1 + R_2} = \frac{600 \times 300}{600 + 300} = 200 \text{ ohms}$$

Calculating $X_{L\ TOT}$: The individual values of X_{L1} and X_{L2} must first be found.

$$X_{L1} = 2\pi fL = 6.28 \times 50 \times 1 = 314 \text{ ohms}$$
$$X_{L2} = 2\pi fL = 6.28 \times 50 \times 2 = 628 \text{ ohms}$$

The two reactances are in parallel, so the product/sum method can also be used to find their total.

$$X_{L\ TOT} = \frac{X_{L1}X_{L2}}{X_{L1} + X_{L2}} = \frac{314 \times 628}{314 + 628} = 209 \text{ ohms}$$

Calculating θ:

$$\tan\theta = \frac{R_{TOT}}{X_{L\ TOT}} = \frac{200}{209} = 0.957$$

θ can now be found from a table of trigonometric functions to be:

$$\theta = 43.7°$$

comparison of series and parallel RL circuits

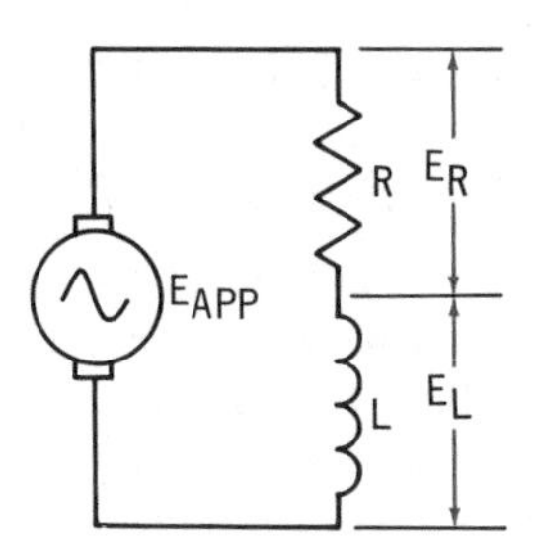

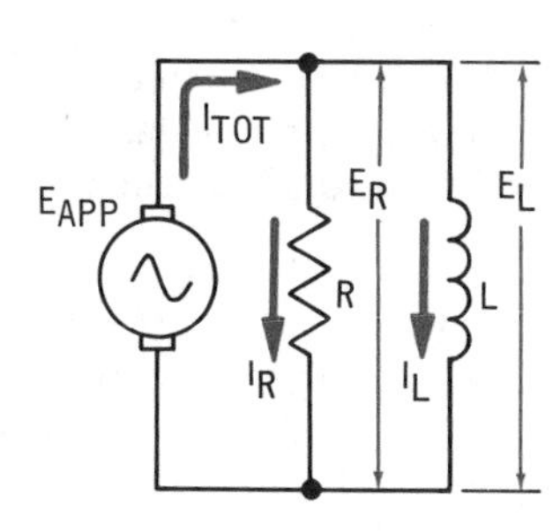

	Series RL Circuit	Parallel RL Circuit
Current	Current is same everywhere in circuit. Current through R and L are, therefore, in phase.	Current divides between resistive and inductive branches. $I_{TOT} = \sqrt{I_R^2 + I_L^2}$ $I_R = E_{APP}/R$ $I_L = E_{APP}/X_L$ Current through R leads current through L by 90°.
Voltage	Vector sum of voltage drops across R and L equals applied voltage. $E_{APP} = \sqrt{E_R^2 + E_L^2}$ Voltage across L leads voltage across R by 90°.	Voltage across each branch is same as applied voltage. Voltages across R and L are, therefore, in phase. $E_R = E_L = E_{APP}$
Impedance	It is the vector sum of resistance and inductive reactance. $Z = \sqrt{R^2 + X_L^2}$	It is calculated in same way as parallel resistances, except that vector addition is used. $Z = RX_L/\sqrt{R^2 + X_L^2}$
Phase Angle (θ)	It is the angle between circuit current and applied voltage. $\tan\theta = E_L/E_R = X_L/R$ $\cos\theta = R/Z$	It is the angle between applied voltage and line current. $\tan\theta = I_L/I_R = R/X_L$ $\cos\theta = Z/R$
Power	Power delivered by source is apparent power. Power actually consumed in circuit is true power. Power factor determines what portion of apparent power is true power. $P_{APPARENT} = E_{APP}I$ $P_{TRUE} = E_{APP}I\cos\theta$ $P.F. = \cos\theta$	
Effect of Increasing Frequency	X_L increases, which in turn causes circuit current to decrease. Phase angle increases, which means that circuit is more inductive.	X_L increases, inductive branch current decreases, and so line current also decreases. Phase angle decreases, which means that circuit is more resistive.
Effect of Increasing Resistance	Phase angle decreases, which means that circuit is more resistive.	Phase angle increases, which means that circuit is more inductive.
Effect of Increasing Inductance	Phase angle increases, which means that circuit is more inductive.	Phase angle decreases, which means that circuit is more resistive.

summary

□ In a simple parallel RL circuit, there is one resistive branch and one inductive branch. □ The circuit current in a parallel RL circuit divides before entering the branches, and a portion flows through the resistive branch, while the rest flows through the inductive branch. □ The current through each branch is independent of the current in the other. □ Both branches are connected directly across the voltage source. □ Voltage is used as the zero reference vector since it is the common circuit quantity in a parallel RL circuit.

□ The current in the resistive branch of a parallel RL circuit is found by $I_R = E/R$. □ The current in the inductive branch is found by $I_L = E/X_L$. □ The line current is found by adding the branch currents vectorially: $I_{LINE} = \sqrt{I_R^2 + I_L^2}$. □ The line current can also be found by: $I_{LINE} = E/Z$. □ The phase angle of a parallel RL circuit is found by: $\tan \theta = I_L/I_R$; or $\tan \theta = R/X_L$; or $\cos \theta = Z/R$. □ The impedance of a parallel RL circuit is found by adding the parallel oppositions vectorially. The equation for impedance is $Z = RX_L/(\sqrt{R^2 + X_L^2})$. □ The impedance can also be found by: $Z = E_{APP}/I_{LINE}$.

□ The power equations for parallel RL circuits are identical to those for series RL circuits. □ Since inductive reactance increases with an increase in frequency, the parallel RL circuit becomes more resistive as the frequency is increased.

review questions

1. What is meant by *line current* in a parallel RL circuit?

For Questions 2 to 10, consider a parallel RL circuit with an applied voltage of 100 volts, a resistor with a resistance of 10 ohms, and an inductor with an inductive reactance of 20 ohms.

2. What is the current through the resistor; inductor?
3. What is the voltage across the resistor; inductor?
4. What is the impedance of the circuit?
5. What is the phase angle of the circuit?
6. What is the apparent power? The true power? The power factor?
7. What is the value of the line current?
8. By how much does the frequency have to be multiplied for the parallel circuit to become, effectively, a resistive circuit?
9. Answer Questions 3 to 5 for the condition where the frequency is doubled.
10. Answer Questions 5 to 7 for the condition where the frequency is tripled.

RC circuits

A circuit that contains *resistance* (R) and *capacitance* (C) is called an *RC circuit.* The methods you use to solve RC circuits depend on whether the resistance and capacitance are in *series* or in *parallel.* This is similar to what you have just learned for RL circuits. Actually, the conditions that exist in RC circuits and the methods used for solving them are quite similar to those for RL circuits. The principal difference is one of *phase relationship,* since as you will remember from Volume 3, the phase relationship between the current and voltage in a capacitive circuit is different from that in an inductive circuit.

RC Circuits

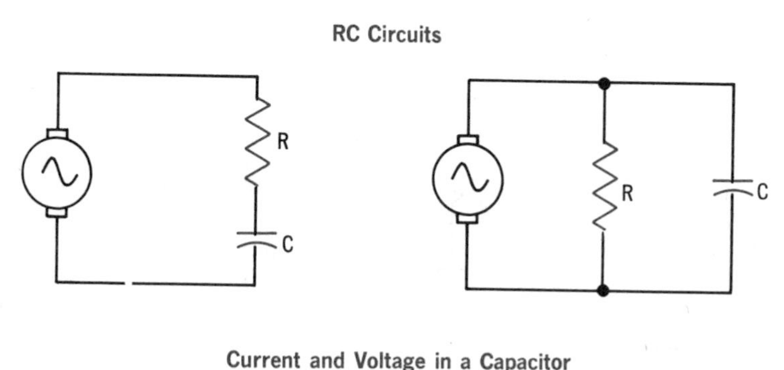

Current and Voltage in a Capacitor

AS WAVEFORMS

+
Current
Voltage
0
–
90°

AS VECTORS

Current
90°
Voltage

RC circuits are series or parallel combinations of resistance and capacitance. The analysis of RC circuits is based on the fact that current in a purely capacitive circuit leads the voltage by 90°

An RC circuit is usually considered as one that contains resistors and capacitors. However, any practical circuit has some resistance in the circuit wiring, as well as some capacitance between wires or between the wiring and surrounding metal parts. RC circuits, therefore, exist even when no resistors or capacitors are used. However, in these cases, the values of resistance and capacitance are usually very small. In this volume, therefore, the resistance and capacitance of the circuit wiring will be considered negligible.

series RC circuits

In a series RC circuit, one or more resistances are connected in series with one or more capacitances, so that *total circuit current* flows through *each* individual component. For the discussion on the following pages of the voltage, impedance, and current in series RC circuits, the case of a *single resistance* in series with a *single capacitance* will be considered, unless otherwise stated. When there is more than one resistance or capacitance, the analysis of the circuit is the same, except that the single resistance or capacitance then becomes the total resistance or capacitance.

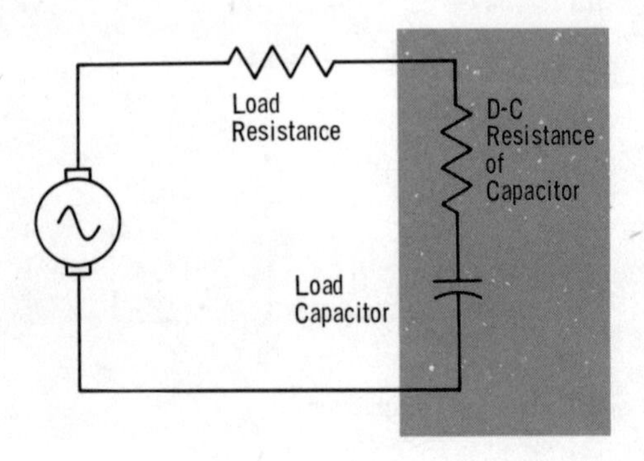

Every capacitor has leakage, which is caused by the d-c resistance of the capacitor. Normally, a capacitor's d-c resistance is very high, so, essentially, the capacitor acts as an ideal capacitor, passing ac and blocking dc

You will recall from Volume 3 that every capacitor has some *leakage*, made up of a small amount of current that flows through the dielectric. Effectively, the leakage current destroys the 90-degree relationship between the voltage across the capacitor and the current through it, so that the current actually leads the voltage by some phase angle less than 90 degrees. For most capacitors, though, the leakage current is so small that for all practical purposes the phase angle can be considered as being 90 degrees. In this volume, therefore, we will consider capacitors as having no leakage, and the capacitor current as leading the voltage by 90 degrees.

In the description of series RC circuits on the following pages, the circuit current will be used as the phase reference for all other circuit quantities, just as it was for series RL circuits. Again, the choice of current is for convenience, since it is the same in all parts of the circuit. With current used as the reference, the vectors of all quantities that are in phase with the current will have the same direction as the current vector: 0 degrees.

voltage

When current flows in a series RC circuit, the voltage drop across the resistance (E_R) is in phase with the current, while the voltage drop across the capacitance (E_C) lags the current by 90 degrees. Since the current through both is the same, E_R leads E_C by 90 degrees. The amplitudes of the two voltage drops can be calculated by:

$$E_R = IR$$
$$E_C = IX_C$$

Like series RL circuits, the *vector sum* of the voltage drops *equals* the applied voltage. As an equation:

$$E_{APP} = \sqrt{E_R^2 + E_C^2}$$

The relationship between the applied voltage and the voltage drops in a series RC circuit is such that the applied voltage equals the VECTOR SUM of the voltage drops

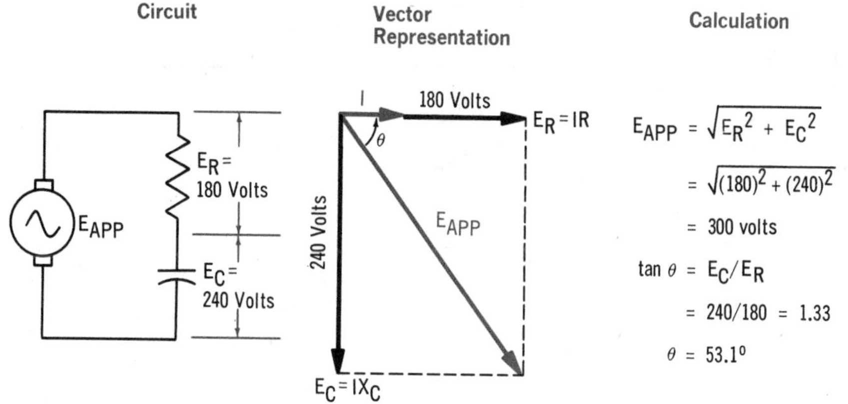

If one of the voltage drops changed as a result of a change in either R or X_C, the angle of the applied voltage vector would also appear to change. Actually, it is the current that changes phase; this is the same as was pointed out for series RL circuits. To avoid confusion, always consider the phase angle, θ, as the angle between I and E_{APP}

Graphically, the applied voltage is the hypotenuse of a right triangle whose two sides are the voltage drops E_R and E_C. The angle (θ) of this vector triangle between the applied voltage and E_R is the same as the phase angle between the applied voltage and the current. The reason for this is that E_R and I are in phase. The value of θ can be calculated from:

$$\tan\theta = E_C/E_R \qquad \text{or} \qquad \cos\theta = E_R/E_{APP}$$

voltage waveforms

The waveforms of the voltages in a series RC circuit are similar to those you have seen for a series RL circuit. They show how the applied voltage waveform is the sum of all the instantaneous points of the two voltage drop waveforms. They also show that the average and effective values of the applied voltage waveform equal the vector sum of the average and effective values of the voltage drop waveforms. This is illustrated for the circuit solved on the previous page.

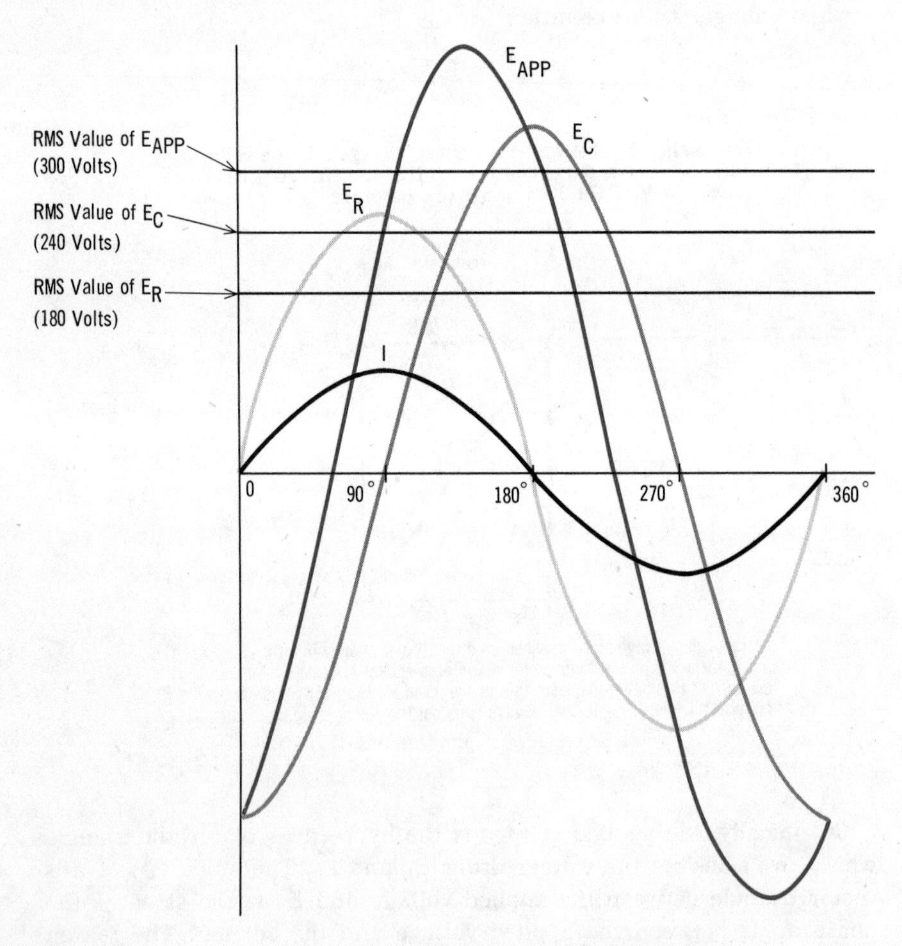

Every point on the applied voltage waveform (E_{APP}) is the algebraic sum of the instantaneous values of the E_R and E_C waveforms

impedance

The impedance of a series RC circuit is the *total* opposition to current flow offered by the circuit resistance and capacitive reactance. It is calculated in the same way as the impedance of a series RL circuit, except that capacitive reactance is used in place of inductive reactance. The equation for the impedance of a series RC circuit is, therefore,

$$Z = \sqrt{R^2 + X_C^2}$$

The vector addition takes into account the 90-degree phase difference between the voltage drop across the resistance and that across the capacitance.

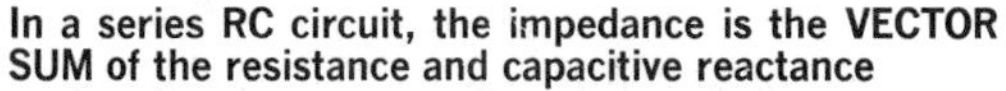

In a series RC circuit, the impedance is the VECTOR SUM of the resistance and capacitive reactance

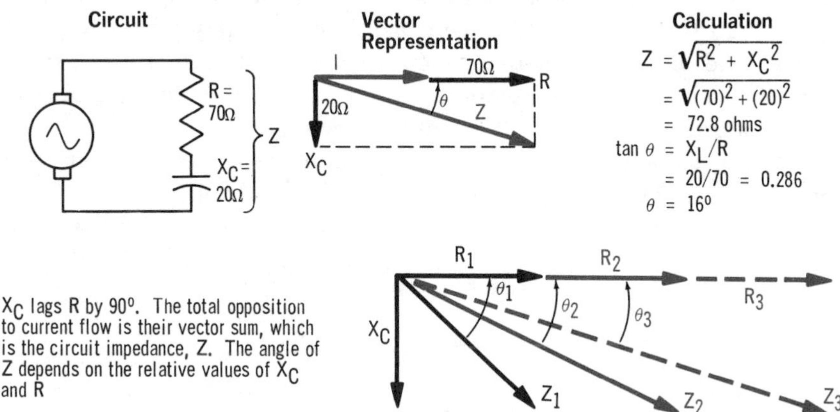

X_C lags R by 90°. The total opposition to current flow is their vector sum, which is the circuit impedance, Z. The angle of Z depends on the relative values of X_C and R

As R becomes larger relative to X_C, the angle of Z becomes smaller

Since R and X_C are 90 degrees apart, with R leading, the phase angle of Z is somewhere between 0 and 90 degrees. The exact angle depends on the *relative* values of R and X_C. If R is greater, Z will be closer to 0 degrees, and if X_C is greater, Z will be closer to 90 degrees. The value of the angle can be calculated from either of the equations:

$$\tan \theta = X_C/R \qquad \text{or} \qquad \cos \theta = R/Z$$

The phase angle of Z is the same as the phase angle between the applied voltage and the current. So if you know one, you automatically know the other.

Just as it does in series RL circuits, the 10-to-1 rule applies to the impedance of series RC circuits. This means that if either X_C or R is 10 or more times greater than the other, the circuit will operate essentially as if only the larger of the two quantities was present.

current

The amplitude of the current in a series RC circuit can be calculated from Ohm's Law if you know the applied voltage and the impedance. Thus,

$$I = E_{APP}/Z \qquad \text{where} \quad Z = \sqrt{R^2 + X_C^2}.$$

Since the current is the *same throughout* the circuit, it is used as the phase reference. So the angle between it and the impedance determines whether the current is more resistive or more capacitive. You will recall that this angle is somewhere between 0 and 90 degrees, with its exact value depending on the *relative* values of the resistance and the capacitive reactance. The *larger* X_C is compared to R, the closer the angle is to *90 degrees*, and the more *capacitive* is the current; similarly, the *smaller* X_C is compared to R, the closer the angle is to *0 degrees*, and the more *resistive* is the current. If X_C is 10 or more times larger than R, the current can be considered as purely capacitive, and thus to lead the applied voltage by 90 degrees; while if R is 10 or more times greater than X_C, you can consider the current as purely resistive, and thus being in phase with the applied voltage.

You can calculate the circuit power using the same equations you learned for RL circuits.

In a series RC circuit, the current is the same at every point, and leads the applied voltage by an angle between 0 and 90°

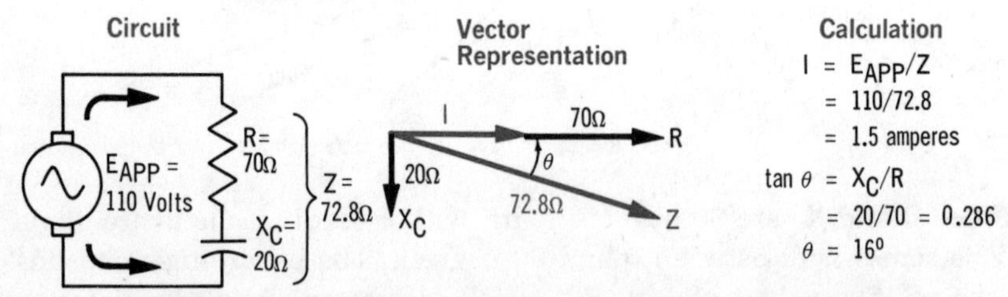

The angle calculated here is the phase angle of the current. You will notice that it is the same as the angle of the impedance found for the same circuit on the previous page. The reason for this is that the value of X_C and R determine the angle of the impedance, which in turn determines if the current is capacitive or resistive

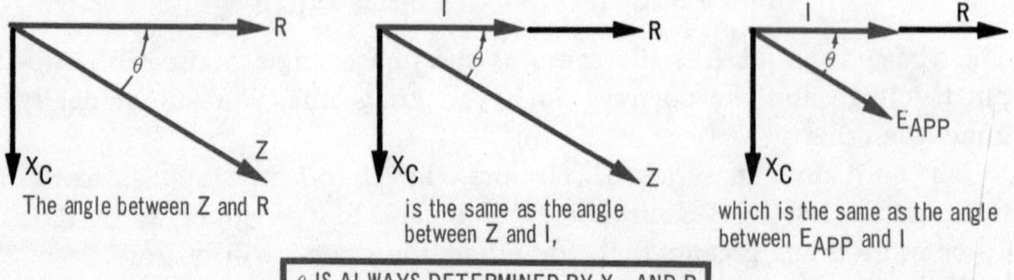

The angle between Z and R is the same as the angle between Z and I, which is the same as the angle between E_{APP} and I

θ IS ALWAYS DETERMINED BY X_C AND R

The characteristics of a series RC circuit vary with different frequencies

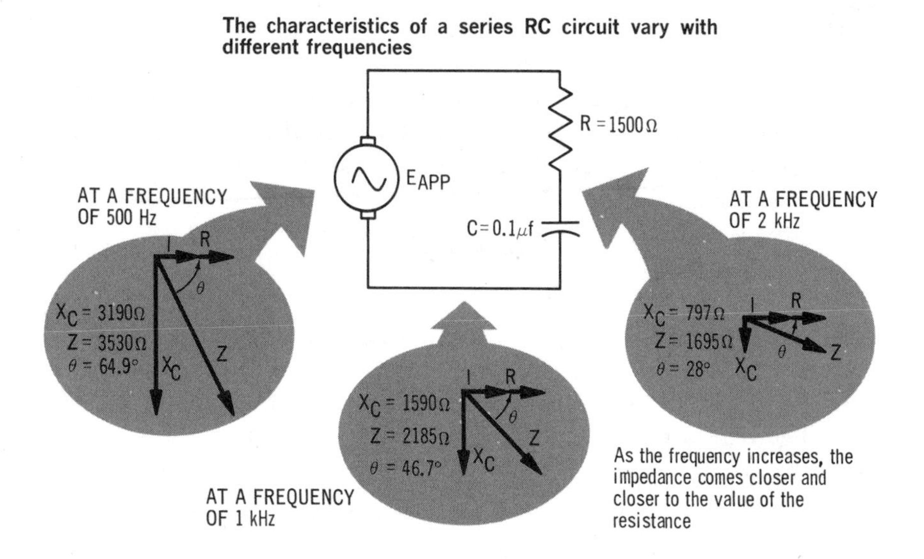

As the frequency increases, the impedance comes closer and closer to the value of the resistance

effect of frequency

Since the value of X_C in a series RC circuit *changes* with *frequency,* all the properties of the circuit that are affected by X_C also change with frequency. These frequency-dependent properties include the *impedance,* the amplitude and phase angle of the *current,* and the circuit *power factor.* Since the value of X_C is inversely proportional to frequency, an increase in frequency causes a decrease in X_C, while a decrease in frequency causes X_C to increase. As a result, when the frequency goes up, the impedance decreases, the circuit current increases and becomes more resistive, and the power factor goes closer to 1. Conversely, when frequency goes down, the impedance increases, the circuit current decreases and becomes more capacitive, and the power factor goes closer to 0.

This is the frequency response curve of the circuit. At very low frequencies, Z is practically infinite; and the higher the frequency becomes, the lower is the impedance, approaching but never actually reaching the value of R

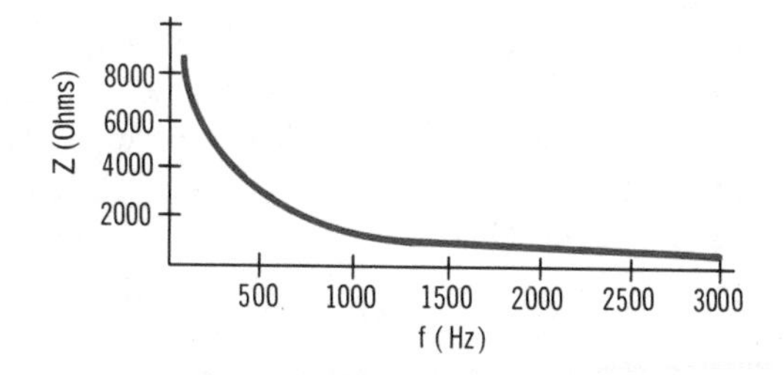

solved problems

What is the applied voltage in the circuit below?

The applied voltage is the unknown quantity that has to be found. Therefore, the first step is to consider the equations for calculating the applied voltage. These equations are

$$E_{APP} = IZ$$

and

$$E_{APP} = \sqrt{E_R^2 + E_C^2}$$

From the information given in the circuit diagram below, it should be obvious that the equation $E_{APP} = IZ$ cannot be used, since to find Z, you must know the value of X_C. In turn, to find the value of X_C, you must know the frequency of the applied voltage, and that information is not given.

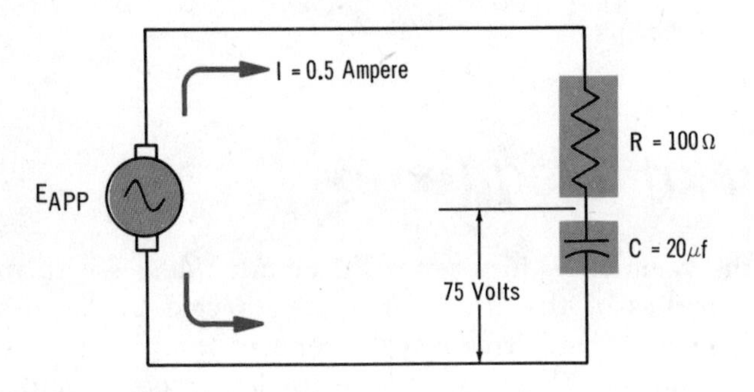

The applied voltage can only be found, therefore, by the equation $E_{APP} = \sqrt{E_R^2 + E_C^2}$. Before this can be done, however, the voltage drop across the resistance has to be calculated.

Calculating E_R:

$$E_R = IR = 0.5 \text{ ampere} \times 100 \text{ ohms} = 50 \text{ volts}$$

Calculating E_{APP}:

$$E_{APP} = \sqrt{E_R^2 + E_C^2} = \sqrt{(50)^2 + (75)^2} = \sqrt{8125} = 90 \text{ volts}$$

Problem 13. What is the impedance of the circuit?

Before the applied voltage was calculated, the impedance could not be determined. The reasons for this were that in the equation for impedance, $Z = E_{APP}/I$, the value of E_{APP} was not known, while in the equation $Z = \sqrt{R^2 + X_C^2}$, the value of X_C was not known. Now that the applied voltage, E_{APP}, has been calculated, however, the impedance can easily be found:

$$Z = E_{APP}/I = 90 \text{ volts}/0.5 \text{ ampere} = 180 \text{ ohms}$$

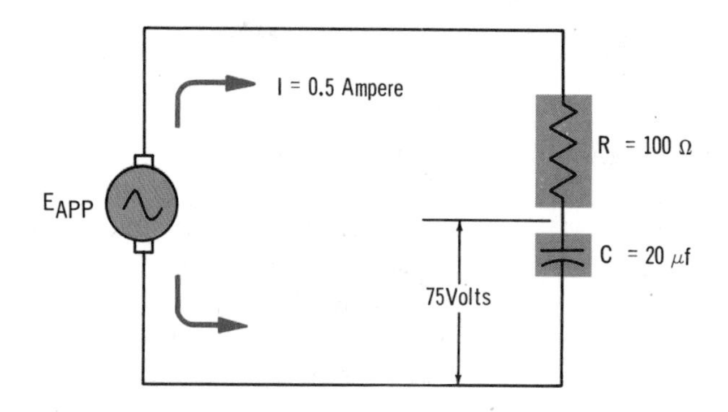

solved problems (cont.)

Problem 14. What is the phase angle in the circuit?

As you know, the three commonly used equations for calculating the phase angle, θ, are

$$\cos\theta = R/Z \qquad \tan\theta = X_C/R \qquad \tan\theta = E_C/E_R$$

The second equation cannot be used in this problem since the value of X_C is not known. Enough information is known, however, to use either of the other two equations.

$$\cos\theta = R/Z = 100/180 = 0.555$$

$$\theta = 56.3°$$

$$\tan\theta = E_C/E_R = 75/50 = 1.500$$

$$\theta = 56.3°$$

What is the frequency of the power source in the circuit?

The only way the frequency can be found is if the value of X_C is known. This can be calculated from the equation for the voltage drop across C, since both I and E_C are known. Thus,

$$E_C = IX_C$$

so,

$$X_C = E_C/I = 75/0.500 = 150 \text{ ohms}$$

With the value of X_C now known, the frequency can be found by the equation for calculating capacitive reactance:

$$X_C = \frac{1}{2\pi fC}$$

so,

$$f = \frac{1}{2\pi CX_C} = \frac{1}{6.28 \times 0.00002 \times 150} = 53 \text{ Hz}$$

solved problems (cont.)

Problem 16. What capacitance should the capacitor have if the lamp is to dissipate its rated wattage?

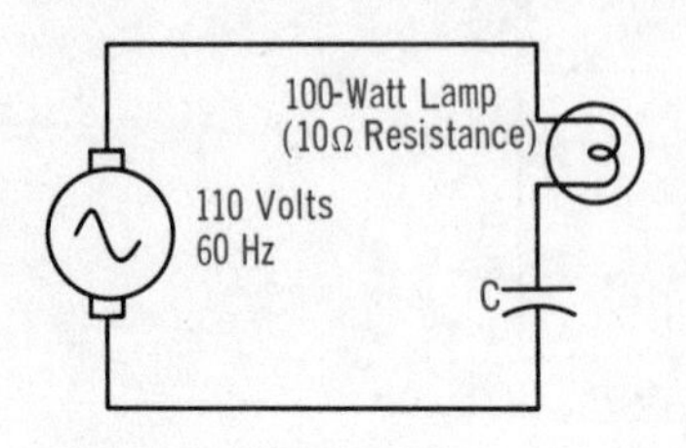

To solve this problem, you will find it convenient to change the basic equations to new forms.

Since you know the lamp's wattage rating and resistance, you first have to find the current that will cause the lamp to dissipate its rated wattage. You start with the equation that relates power (P), current (I), and resistance (R), or, $P = I^2R$. This equation can be changed to a form that will allow you to solve for the current as follows:

$$P = I^2R \quad \text{or} \quad I^2 = P/R, \text{ which becomes } I = \sqrt{P/R}$$

Now you can calculate the current required by the lamp:

$$I = \sqrt{P/R} = \sqrt{100/10} = \sqrt{10} = 3.16 \text{ amperes}$$

It is a series circuit, so 3.16 amperes must flow through the entire circuit. With the applied voltage of 110 volts, the circuit impedance that will allow 3.16 amperes to flow can be found by Ohm's Law in the form:

$$Z = E_{APP}/I = 110 \text{ volts}/3.16 \text{ amperes} = 34.8 \text{ ohms}$$

The circuit impedance, which is the vector sum of the resistance of the lamp and the reactance of the capacitor, must therefore be 34.8 ohms. You know the lamp resistance (R) and the circuit impedance (Z). To find the capacitive reactance (X_C), the equation for impedance, $Z = \sqrt{R^2 + X_C^2}$, can be changed as follows:

$$Z = \sqrt{R^2 + X_C^2} \text{ which becomes } Z^2 = R^2 + X_C^2$$

Transposing, $X_C^2 = Z^2 - R^2$ which becomes $X_C = \sqrt{Z^2 - R^2}$

Therefore, $X_C = \sqrt{(34.8)^2 - (10)^2} = \sqrt{1111} = 33.3$ ohms

The capacitor must thus have a capacitive reactance of 33.3 ohms. To find the value of capacitance that has a reactance of 33.3 ohms with a 60-Hz applied voltage, you start with the basic equation for capacitive reactance and change it as follows:

$$X_C = \frac{1}{2\pi fC} \quad \text{or} \quad 2\pi fCX_C = 1$$

Solving for C:

$$C = \frac{1}{2\pi fX_C}$$

Therefore,

$$C = \frac{1}{6.28 \times 60 \times 33.3} = \frac{1}{12{,}547} = 79.7 \text{ microfarads}$$

summary

□ In a series RC circuit, since the circuit current flows through both the resistance and the capacitance, it is used as the phase reference. □ The voltage drop across the resistance is in phase with the current, while the voltage drop across the capacitance lags the current by 90 degrees. □ The voltage drop across the resistance can be found by $E_R = IR$. □ The voltage drop across the capacitance can be found by $E_C = IX_C$.

□ The vector sum of the voltage drops around a series RC circuit equals the applied voltage: $E_{APP} = \sqrt{E_R^2 + E_C^2}$. □ The phase angle can be found by: $\tan\theta = E_C/E_R$; or $\tan\theta = X_C/R$; or $\cos\theta = E_R/E_{APP}$; or $\cos\theta = R/Z$. □ The equation for impedance in a series RC circuit is $Z = \sqrt{R^2 + X_C^2}$. □ The current is the same through the circuit, and is given by: $I = E_{APP}/Z$.

□ The power equations for a series RC circuit are similar to those for series RL circuits. □ Since the capacitive reactance, X_C, decreases with an increase in frequency, as the frequency increases, a series RC circuit becomes more resistive.

review questions

1. On a vector diagram of the voltages in a series RC circuit, what circuit quantity is taken as the reference vector? Why?
2. What is the 10-to-1 rule for resistance and capacitive reactance for a series RC circuit?
3. What is meant by *leakage current*? What effect does leakage current have on the phase relationship for the voltage and current of a capacitor?
4. What is the resistance in a series RC circuit when the impedance is 130 ohms, and the capacitive reactance is 50 ohms?
5. What is the phase angle for the circuit of Question 4?
6. The applied voltage across a series RC circuit is 100 volts, and causes a current of 5 amperes to flow. What is the magnitude of the impedance of the current?
7. The power dissipated in a circuit is 500 watts, the impedance is 10 ohms, and the phase angle is 60 degrees. What is the value of the current in the circuit?
8. What is the value of the apparent power of a circuit that dissipates 500 watts, and has a power factor of 0.25?
9. Between what values can the power factor of a series RC circuit be found? Why?
10. What is the voltage across the capacitor in a series RC circuit when the applied voltage is 100 volts, and the voltage across the resistor is 80 volts?

parallel RC circuits

In a parallel RC circuit, one or more resistive loads and one or more capacitive loads are connected in parallel across a voltage source. There are, therefore, *resistive branches,* containing only resistance, and *capacitive branches,* containing only capacitance. The current that leaves the voltage source divides among the branches, so there are different currents in different branches. The current is, therefore, not a common quantity, as it is in series RC circuits.

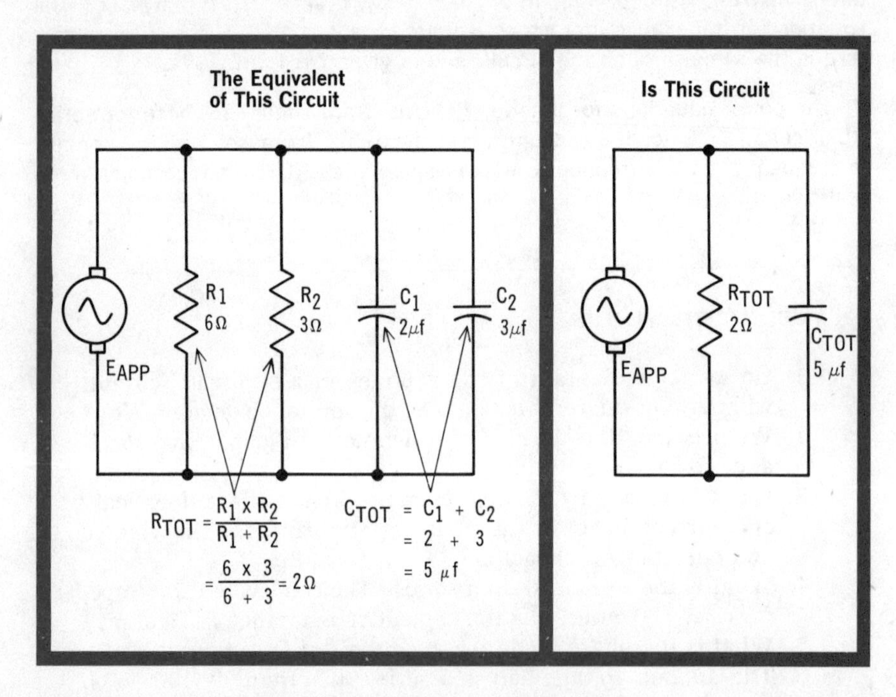

When calculating the total circuit quantities of applied voltage, line current, impedance, and power, resistive and capacitive branches should first be reduced to their simple single equivalents

The description of parallel RC circuits contained on the following pages will cover parallel circuits that contain only a single resistive branch and a single capacitive branch. Circuits that contain more than one resistive or capacitive branch are identical, except that when solving them for overall circuit quantities such as impedance or line current, the resistive or capacitive branches must first be reduced to their equivalent single resistive or capacitive branch.

voltage

In a parallel RC circuit, as in any parallel circuit, the applied voltage is *directly across* each branch. The branch voltages are, therefore, equal to each other, as well as to the applied voltage, and all three are *in phase*. So if you know any one of the circuit voltages, you know all of them. Since the voltage is *common* throughout the circuit, it serves as the common quantity in any vector representation of parallel RC circuits. This means that on any vector diagram, the reference vector will have the same direction, or phase relationship, as the circuit voltage. The two quantities that have this relationship with the circuit voltage, and whose vectors therefore have a direction of zero degrees, are the circuit resistance and the current through the resistance.

In a parallel RC circuit, each branch voltage is the same as the applied voltage

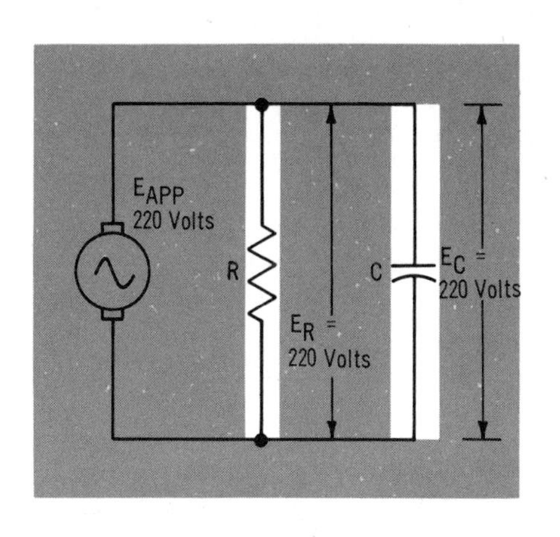

Phase relationships between the various quantities in a parallel RC circuit are expressed in relation to how they differ phasewise from the circuit voltage. The reason for this is that the voltage is the same throughout the circuit, and so provides a basis for expressing phase differences

As a Vector

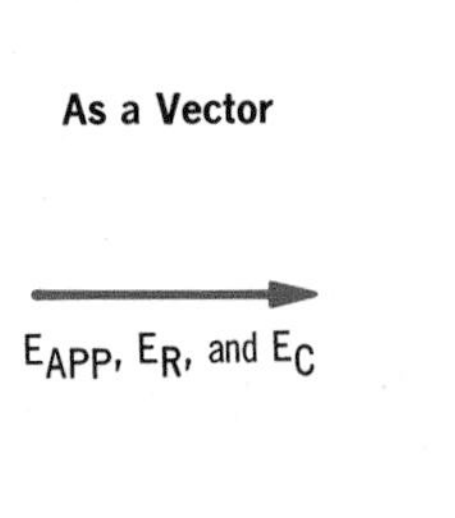

As a Waveform

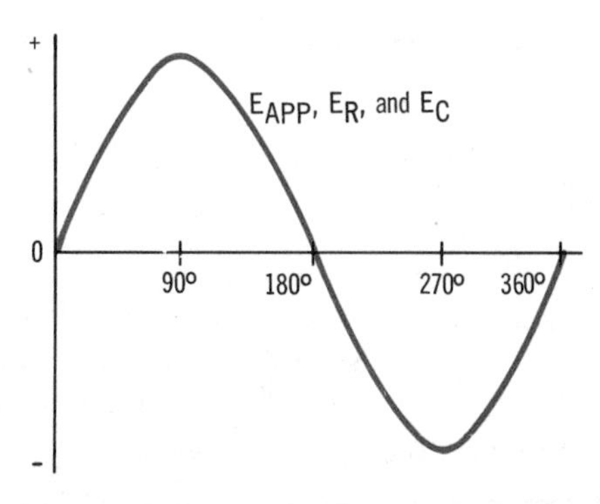

branch current

The current in each branch of a parallel RC circuit is *independent* of the current in the other branches. Current within a branch depends only on the voltage across the branch, and the resistance or capacitive

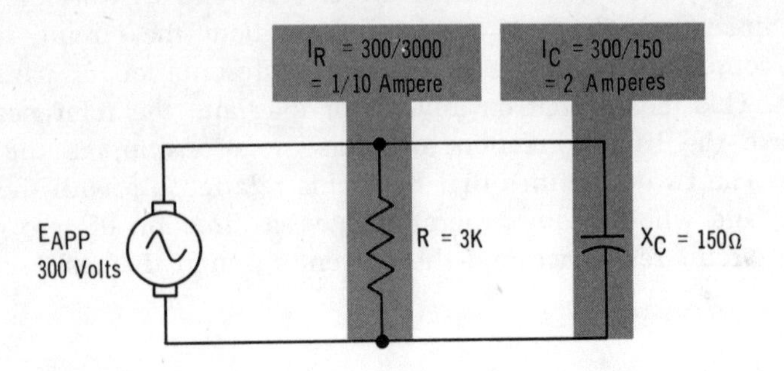

reactance contained in it. The current in the resistive branch is calculated from the equation:

$$I_R = E_{APP}/R$$

The current in the capacitive branch is found with the equation:

$$I_C = E_{APP}/X_C$$

The current in the resistive branch is in phase with the branch voltage,

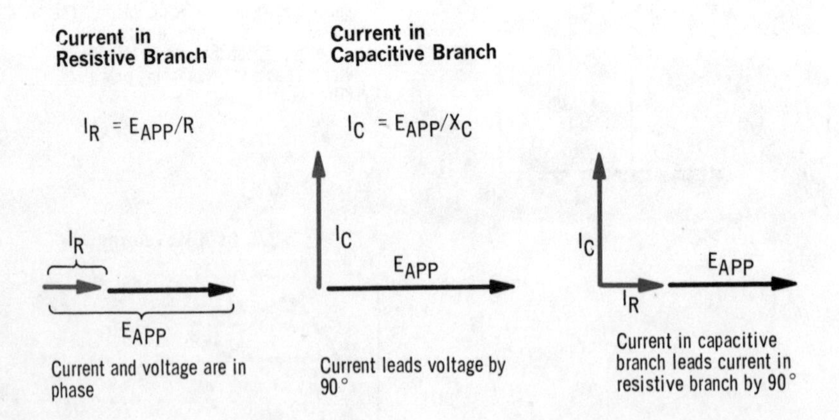

while the current in the capacitive branch leads the branch voltage by 90 degrees. Since the two branch voltages are the same, the current in the *capacitive* branch (I_C) must *lead* the current in the *resistive* branch (I_R) by 90 degrees.

Line current in a parallel RC circuit is equal to the VECTOR SUM of the currents in the resistive and capacitive branches

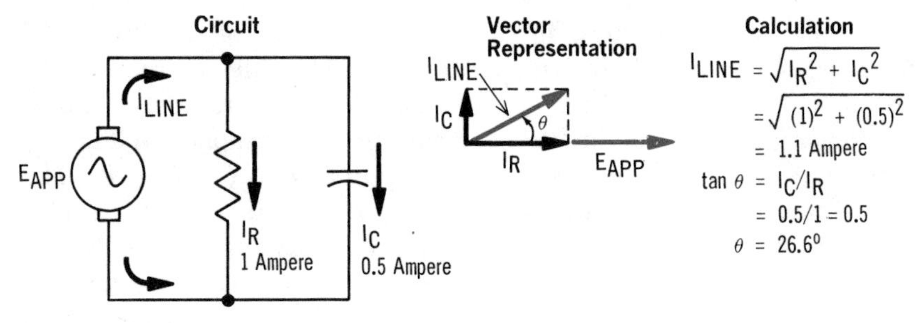

line current

Since the branch currents in a parallel RC circuit are out of phase with each other, they have to be added *vectorially* to find the line current. The two branches are 90 degrees out of phase, so their vectors form a right triangle, whose hypotenuse is the line current. The equation for calculating the line current, I_{LINE}, is

$$I_{LINE} = \sqrt{I_R^2 + I_C^2}$$

If the impedance of the circuit and the applied voltage are known, the line current can also be calculated from Ohm's Law.

$$I_{LINE} = E/Z$$

The line current is the vector sum of the resistive and capacitive currents

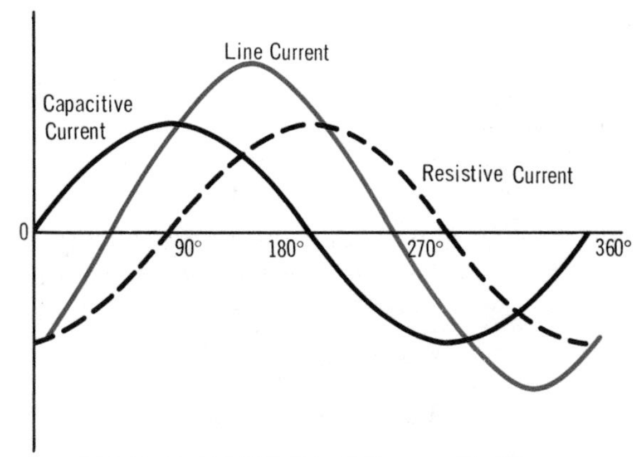

Capacitive current leads the resistive current by 90°

line current (cont.)

Inasmuch as the current in the resistive branch of a parallel RC circuit is in phase with the applied voltage, while the current in the capacitive branch leads the applied voltage by 90 degrees, the sum of the two branch currents, or line current, leads the applied voltage by some phase angle less than 90 degrees but greater than 0 degrees. The exact angle depends on whether the capacitive current or resistive current is greater. If there is more capacitive current, the angle will be closer to 90 degrees; while if the resistive current is greater, the angle is closer to 0 degrees. In cases where one of the currents is 10

The line current in a parallel RC circuit will have a phase angle somewhere between 0 and 90°, leading. The value of θ depends on the relative values of the capacitive and resistive currents in the branches

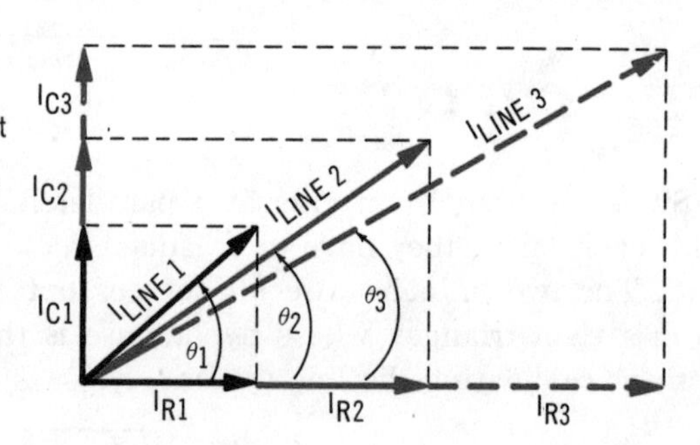

or more times greater than the other, the line current can be considered to have a phase angle of 0 degrees if the resistive current is the larger, or 90 degrees if the capacitive current is the larger. The value of the phase angle can be calculated from the values of the two branch currents with the equation:

$$\tan \theta = I_C/I_R$$

By substituting the quantities $I_C = E/X_C$ and $I_R = E/R$ in the above equation, two other useful equations for calculating the phase angle, θ, can be derived. They are:

$$\tan \theta = R/X_C \qquad \cos \theta = Z/R$$

Once you know the line current and the applied voltage in a parallel RC circuit, you can find the circuit power using the same equations you learned for parallel RL circuits. These are:

$$P_{APPARENT} = E_{APP} I_{LINE}$$

$$P_{TRUE} = E_{APP} I_{LINE} \cos \theta$$

where $\cos \theta$ is the power factor.

current waveforms

Since the branch currents in a parallel RC circuit are out of phase, their *vector sum* rather than their arithmetic sum equals the line current. This is the same condition that exists for the voltage drops in a series RC circuit. By adding the currents vectorially, you are adding their *instantaneous* values at every point, and then finding the average or effective value of the resulting current. This can be seen from the current waveforms shown. They are the waveforms for the circuit solved on page 4-69.

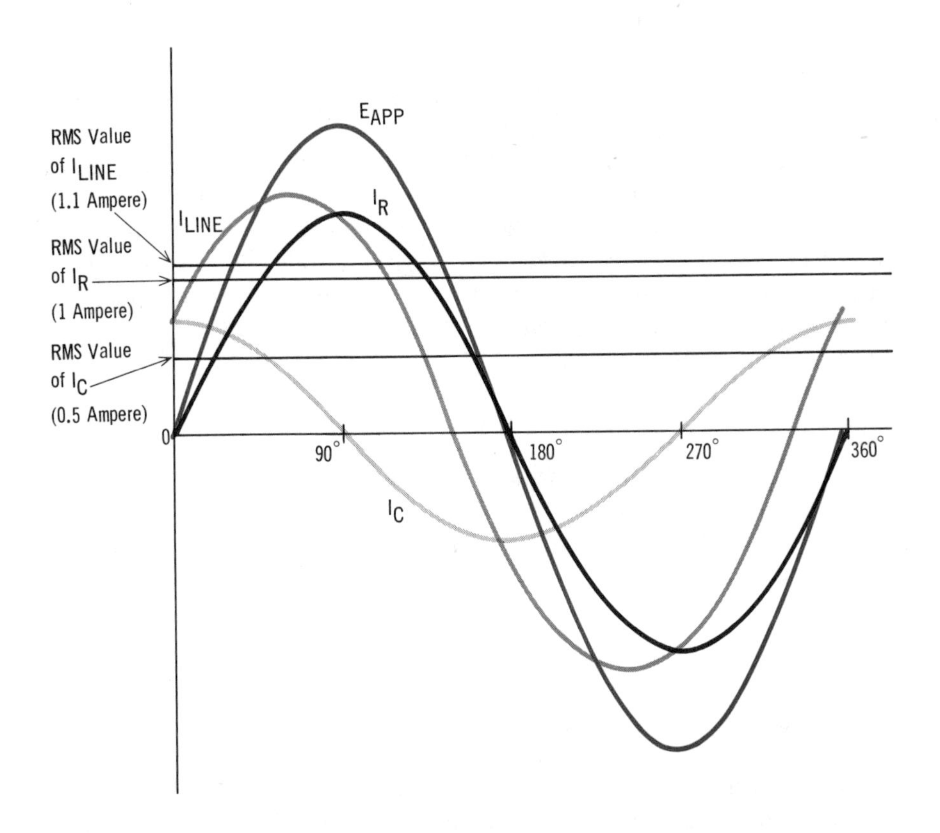

Every point on the line current waveform (I_{LINE}) is the algebraic sum of the instantaneous values of the I_R and I_C waveforms. The rms value of the line current waveform is thus shown to be equal to the vector sum of the rms values of the I_R and I_C waveforms.

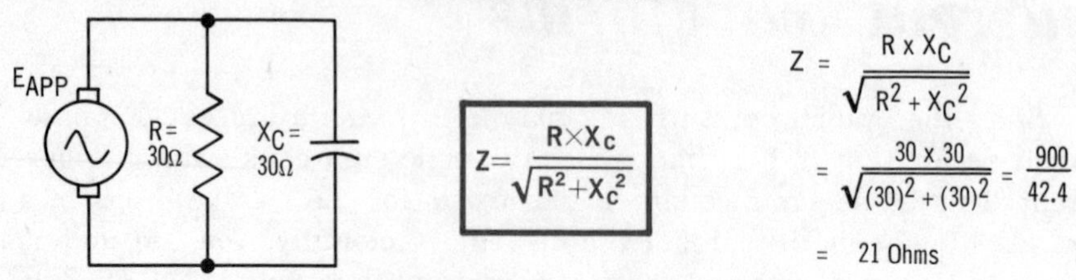

When resistance and capacitive reactance are equal, the impedance is not one-half the value of either one, or 15 ohms, as you might suppose based on your knowledge of parallel resistances

impedance

The impedance of a parallel RC circuit represents the total opposition to current flow offered by the resistance of the resistive branch and the capacitive reactance of the capacitive branch. Like the impedance of a parallel RL circuit, it can be calculated with an equation that is similar to the one used for finding the total resistance of two parallel resistances. However, just as you learned for parallel RL circuits, two vector quantities cannot be added *directly; vector addition* must be used. Therefore, the equation for calculating the impedance of a parallel RC circuit is

$$Z = \frac{RX_C}{\sqrt{R^2 + X_C^2}}$$

where $\sqrt{R^2 + X_C^2}$ is the vector addition of the resistance and capacitive reactance.

In cases where you know the applied voltage and the circuit line current, the impedance can be found simply by using Ohm's Law in the form:

$$Z = E_{APP}/I_{LINE}$$

The impedance of a parallel RC circuit is always *less* than the resistance or capacitive reactance of the individual branches.

The relative values of X_C and R determine how capacitive or resistive the circuit line current is. The one that is the *smallest*, and therefore allows *more* branch current to flow, is the determining factor. Thus, if X_C is smaller than R, the current in the capacitive branch is larger than the current in the resistive branch, and the line current tends to be more capacitive. The opposite is true if R is smaller than X_C. When X_C or R is 10 or more times greater than the other, the circuit will operate for all practical purposes as if the branch with the larger of the two did not exist.

effect of frequency

As in all RL and RC circuits, the frequency of the applied voltage determines many of the characteristics of a parallel RC circuit. Frequency affects the value of the capacitive reactance, and so also affects the circuit impedance, line current, and phase angle, since they are determined to some extent by the value of X_C. The higher the frequency of a parallel RC circuit, the lower is the value of X_C. This means that for a given value of R, the impedance is also lower, making the line current larger and more capacitive. Conversely, the lower the frequency, the greater is the value of X_C, the larger is the impedance, and the

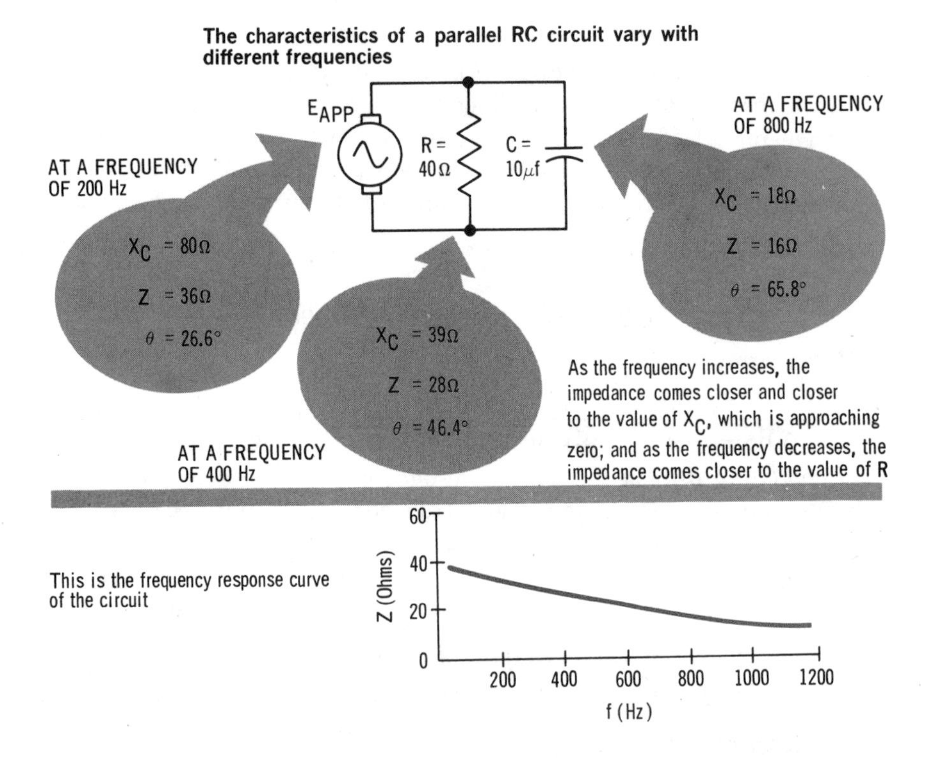

smaller and more resistive is the line current. The same circuit, therefore, can have a *small, resistive* line current at *low frequencies,* and a *large, capacitive* line current at *high frequencies.* Furthermore, since according to the 10-to-1 rule, you can neglect the branch with the smaller current when one of them is 10 or more times larger than the other, depending on the frequency, a parallel RC circuit can act as a simple *series* resistive or capacitive circuit.

solved problems

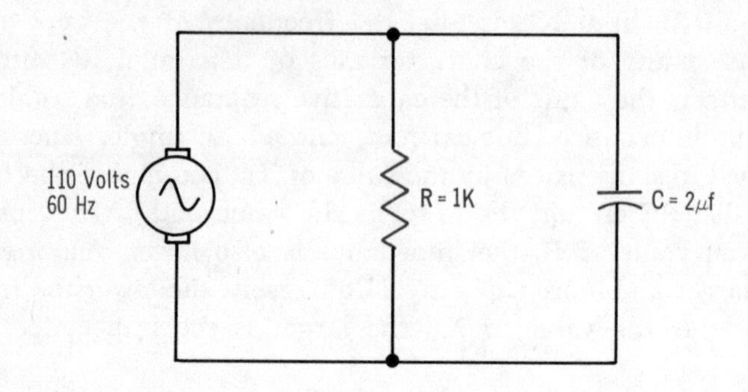

Problem 17. In the circuit, find the impedance, the branch currents, the line current, the phase angle (θ), and the true power.

An inspection of the circuit diagram shows that you are being asked to solve for all of the unknown quantities, with the exception of the capacitive reactance, X_C, and the apparent power. However, to find the impedance, you have to know X_C, and so it too must be calculated.

Calculating X_C:

$$X_C = \frac{1}{2\pi fC} = \frac{1}{6.28 \times 60 \times 0.000002} = \frac{1}{0.000754} = 1326 \text{ ohms}$$

Calculating Z:

$$Z = \frac{RX_C}{\sqrt{R^2 + X_C^2}} = \frac{1000 \times 1326}{\sqrt{(1000)^2 + (1326)^2}} = \frac{1,326,000}{1661} = 798 \text{ ohms}$$

Calculating Branch Currents: You should be aware here that to find the line current, phase angle, and true power it is not necessary that you calculate the branch currents, since you know the applied voltage and impedance, and could therefore use Ohm's Law. The branch currents are being calculated only because they were asked for.

$$I_R = E/R = 110 \text{ volts}/1000 \text{ ohms} = 0.11 \text{ ampere}$$

$$I_C = E/X_C = 110 \text{ volts}/1326 \text{ ohms} = 0.083 \text{ ampere}$$

Calculating I_{LINE}:

$$I_{LINE} = \sqrt{I_R^2 + I_C^2} = \sqrt{(0.11)^2 + (0.083)^2} = \sqrt{0.019} = 0.14 \text{ ampere}$$

Calculating θ:

$$\tan \theta = R/X_C = 1000/1326 = 0.754 \qquad \theta = 37°$$

Calculating True Power:

$$P_{TRUE} = EI_{LINE} \cos \theta$$

$$= 110 \times 0.14 \times 0.798 = 12.3 \text{ watts}$$

comparison of series and parallel RC circuits

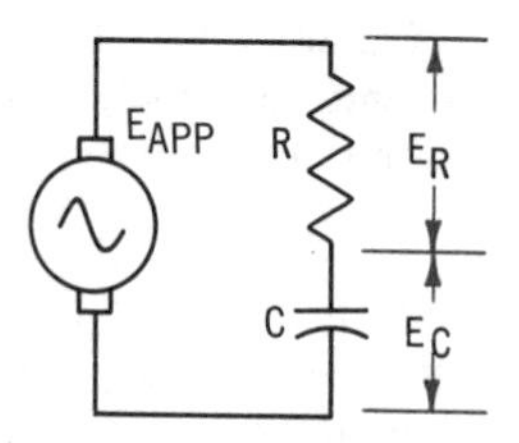

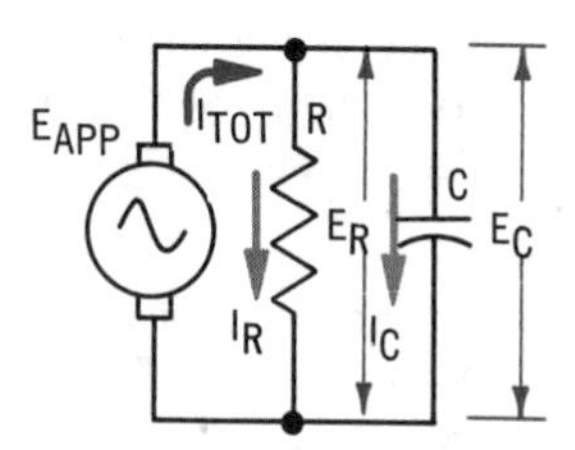

	Series RC Circuit	Parallel RC Circuit
Current	It is the same everywhere in circuit. Currents through R and C are, therefore, in phase.	It divides between resistive and capacitive branches. $I_{TOT} = \sqrt{I_R^2 + I_C^2}$ $I_R = E_{APP}/R$ $I_C = E_{APP}/X_C$ Current through C leads current through R by 90°.
Voltage	Vector sum of voltage drops across R and C equals applied voltage. $E_{APP} = \sqrt{E_R^2 + E_C^2}$ Voltage across C lags voltage across R by 90°.	Voltage across each branch is same as applied voltage. Voltages across R and C are, therefore, in phase. $E_R = E_C = E_{APP}$
Impedance	It is the vector sum of resistance and capacitive reactance. $Z = \sqrt{R^2 + X_C^2}$	It is calculated the same as parallel resistances, except that vector addition is used. $Z = RX_C/\sqrt{R^2 + X_C^2}$
Phase Angle (θ)	It is the angle between circuit current and applied voltage. $\tan\theta = E_C/E_R = X_C/R$ $\cos\theta = R/Z$	It is the angle between line current and applied voltage. $\tan\theta = I_C/I_R = R/X_C$ $\cos\theta = Z/R$
Power	Power delivered by source is apparent power. Power actually consumed in circuit is true power. Power factor determines what portion of apparent power is true power. $P_{APPARENT} = E_{APP}I$ $P_{TRUE} = E_{APP}I\cos\theta$ $P.F. = \cos\theta$	
Effect of Increasing Frequency	X_C decreases, which in turn causes circuit current to increase. Phase angle decreases, which means that circuit is more resistive.	X_C decreases, capacitive branch current increases, and so line current also increases. Phase angle increases, which means that circuit is more capacitive.
Effect of Increasing Resistance	Phase angle decreases, which means that circuit is more resistive.	Phase angle increases, which means that circuit is more capacitive.
Effect of Increasing Capacitance	Phase angle increases, which means that circuit is more capacitive.	Phase angle decreases, which means that circuit is more resistive.

summary

☐ In a parallel RC circuit, the applied voltage is the same across each branch. It is therefore used as the phase reference. ☐ The current through each branch of a parallel RC circuit is independent of the current through the other branches. ☐ The current in the resistive branch is found by $I_R = E_{APP}/R$. ☐ The current in the capacitive branch is found by $I_C = E_{APP}/X_C$. ☐ The line current is found by adding the branch currents vectorially. $I_{LINE} = \sqrt{I_R^2 + I_C^2}$. It can also be found by Ohm's Law for a-c circuits: $I = E/Z$.

☐ The phase angle of a parallel RC circuit is given by: $\tan \theta = I_C/I_R$; or $\tan \theta = R/X_C$; or $\cos \theta = Z/R$. ☐ The impedance is found by adding the parallel oppositions vectorially. The equation for impedance is $Z = RX_C/(\sqrt{R^2 + X_C^2})$.

☐ The power equations for a parallel RC circuit are identical to those for series RC circuits. ☐ Since capacitive reactance, X_C, decreases with increasing frequency, the parallel RC circuit approaches a pure capacitive circuit as the frequency is increased.

review questions

1. What is meant by *line current* in a parallel RC circuit?

For Questions 2 to 10, consider a parallel RC circuit with an applied voltage of 100 volts, a resistor with a resistance of 20 ohms, and a capacitor with a capacitive reactance of 10 ohms.

2. What is the current through the resistor? Through the capacitor?
3. What is the voltage across the resistor? Across the capacitor?
4. What is the impedance of the circuit?
5. What is the phase angle of the circuit?
6. What is the apparent power? The true power? The power factor?
7. What is the value of the line current?
8. By how much does the frequency have to be multiplied for the parallel circuit to become, effectively, a capacitive circuit?
9. Answer Questions 3 to 5 for the condition where the frequency is doubled.
10. Answer Questions 5 to 7 for the condition where the frequency is tripled.

LCR circuits

You have learned the fundamental properties of resistive, inductive, and capacitive circuits, as well as circuits that contain resistance and inductance, and resistance and capacitance. You will now learn about circuits that contain all three of the basic properties of inductance (L), capacitance (C), and resistance (R). These circuits are called *LCR circuits,* and may consist of either series or parallel combinations of inductance, capacitance, and resistance. You will find that everything you have learned previously about resistive, inductive, and capacitive circuits is involved in the analysis of LCR circuits. In addition, though, some entirely new properties and characteristics are involved.

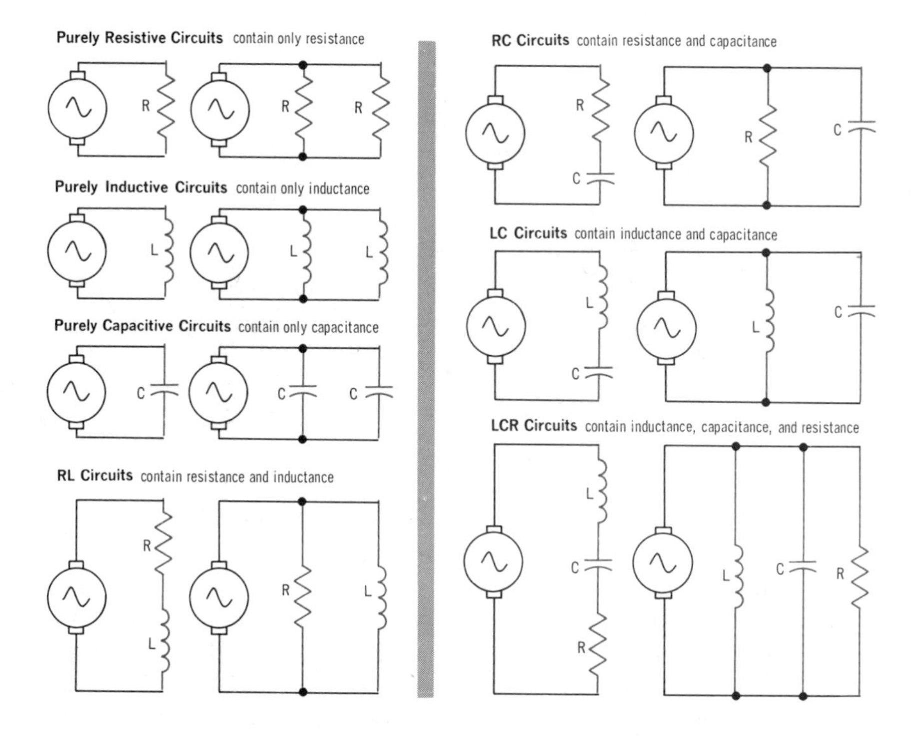

The description of LCR circuits given on the following pages is divided into two parts: one covering series circuits, and the other parallel circuits. For both the series and parallel types, pure *LC circuits* are covered first. These are circuits that have inductance and capacitance, but no resistance. After LC circuits are thoroughly described, resistance will be included, and practical LCR circuits analyzed.

series LC circuits

A series LC circuit consists of an *inductance* and a *capacitance* connected in *series* with a voltage source. There is *no resistance* in the circuit. Of course, this is impossible in actual practice, since every circuit contains some resistance. However, since the circuit resistance of the wiring, the coil winding, and the voltage source is usually so small, it has little or no effect on circuit operation.

A Series LC Circuit

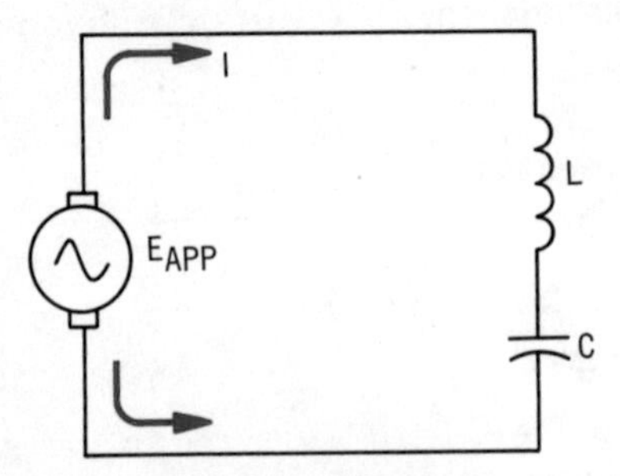

In a series LC circuit, the inductance and capacitance are connected in series, so the total circuit current flows through both. Since the inductance and capacitance are in series, their reactances, X_L and X_C, are also in series

E_{APP} = ?
I =
L =
X_L =
C = ?
X_C =
Z =
θ =

The quantities in series LC circuits that you will normally be interested in are the applied voltage, E_{APP}; the current, I; the inductance, L; the inductive reactance, X_L; the capacitance, C; the capacitive reactance, X_C; the impedence, Z; and the phase angle, θ

As in all series circuits, the *current* in a series LC circuit is the *same* at all points. So the current in the inductance is the same as, and therefore in phase with, the current in the capacitance. Because of this, on any vector diagram of a series LC circuit the direction of the *current vector* is the *reference*, or 0-degree, *direction*. All other quantities, such as the applied voltage and the voltage drops in the circuit, are expressed on the basis of their phase relationship to the circuit current. Here again, as in RL and RC circuits, current is chosen as the phase reference for convenience; not because the phase angle of the current is fixed and does not change. The phase angle of the current depends on the circuit properties, and so it does change. What is important is the *phase difference* between the current and the applied voltage; and vectorially this is the same, whether current or voltage is used as the phase reference.

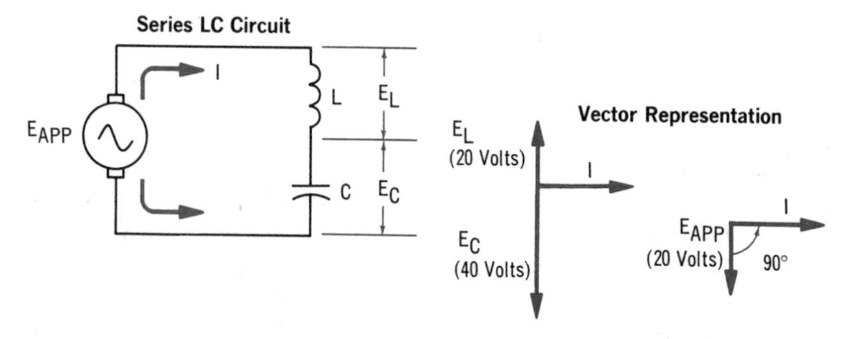

The applied voltage is equal to the vector sum of the voltage drop across the inductance (E_L) and the drop across the capacitance (E_C). If E_L is larger than E_C, the current is purely inductive, so the applied voltage leads the current by 90°; and if E_C is larger than E_L, the current is purely capacitive, so the applied voltage lags the current by 90°

voltage

When a-c current flows in a series LC circuit, the voltages dropped across the inductance and the capacitance depend on the circuit current and the values of X_L and X_C. The voltages can be found by:

$$E_L = IX_L \quad \text{and} \quad E_C = IX_C$$

The voltage across the *inductance leads* the current through it by 90 degrees, while the voltage across the *capacitance lags* the current through it by 90 degrees. Since the *current* through both is the *same,* the voltage across the *inductance leads* that across the *capacitance* by 180 degrees. You will remember that in a series RL or RC circuit, because of the phase differences, the *vector sum* of the voltage drops equals the applied voltage. This is also so in a series LC circuit. There is a difference, though, in how vector addition is used for LC circuits.

In RL and RC circuits, the voltage vectors are 90 degrees out of phase, so the Pythagorean Theorem is used to add them. However, the method used to add vectors that differ by 180 degrees, you remember, is to subtract the smaller vector from the larger and assign the resultant the direction of the larger. When applied to LC series circuits, this means that the applied voltage is equal to the *difference* between the voltage drops across the inductance and capacitance, with the *phase angle* between the applied voltage and the circuit current determined by the *larger* of the voltage drops. As an example, suppose that the voltage drop across the inductance was 50 volts and that across the capacitance 40 volts. The applied voltage would be 10 volts (50 – 40), and would lead the current by 90 degrees, since the larger voltage drop was across the inductance.

voltage (cont.)

A unique property of series LC circuits is that one of the voltage drops, either E_L or E_C, is always *greater* than the *applied voltage*. Moreover, in some cases, both of the voltage drops are greater than the applied voltage. The reason for this, as you will learn more about later, is that the reactances of the inductance and capacitance play a *dual role* in the circuit. They act *together* in opposing the circuit current, whereas they act *independently* in causing their voltage drops.

When the reactances act together, their phase relationship is such that they tend to cancel a portion of each other. The total opposition they offer to the current is therefore less than either would offer individually, so a larger current flows than either would allow by itself. When this current flows, though, it causes a voltage drop across the full value of each individual reactance. In effect, the voltage source "sees" a circuit with little opposition to current flow, and so puts out a relatively large current. But in flowing around the circuit, the current meets opposition which the source does not "see."

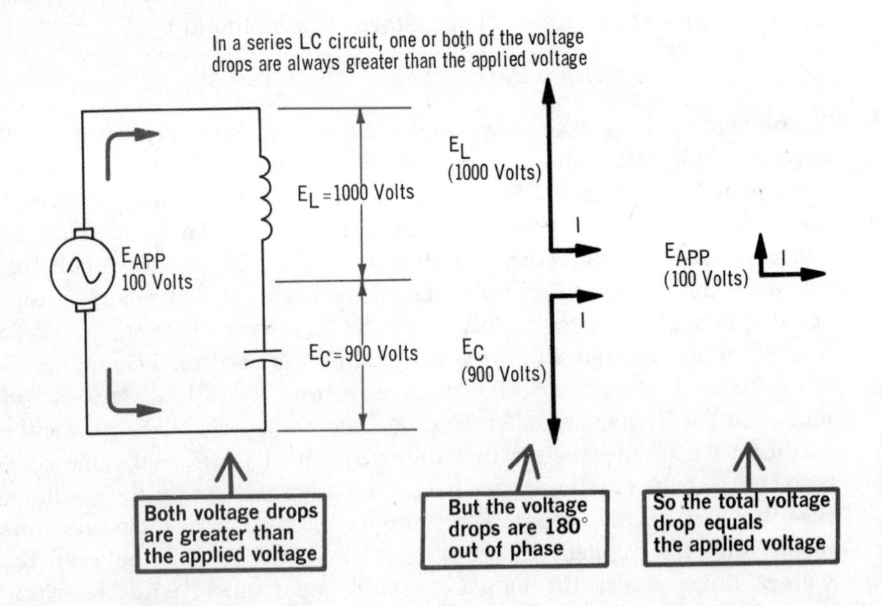

An important point that you should remember is that although one or both of the voltage drops is greater than the applied voltage, they are *180 degrees out of phase*. One of them effectively cancels a portion of the other, so that the total voltage drop is always equal to the applied voltage.

voltage waveforms

You saw on the previous page a circuit that had individual voltage drops of 1000 volts and 900 volts, and yet the voltage source was only 100 volts. The reason for this was that the voltage drops were 180 degrees out of phase. Because of their phase difference, the two voltage drops had to be added vectorially. And since their phase difference was such that one voltage, in effect, *canceled* a portion of the other, the vector addition was done by *algebraic addition.* Thus, one voltage was considered positive, and the other negative, and the two quantities were added algebraically. As shown, the waveforms of the two voltage drops can also be added algebraically, and the average or effective value of the resulting waveform is the average or effective value of the applied voltage.

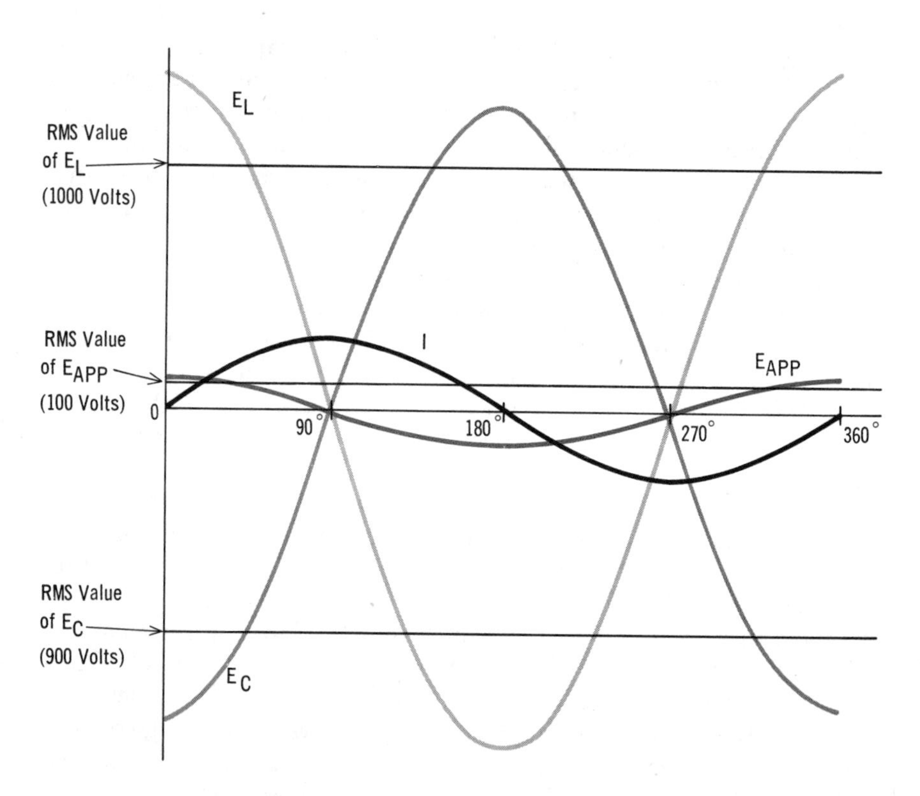

Every point on the applied voltage waveform (E_{APP}) is the algebraic sum of the instantaneous values of the E_L and E_C waveforms

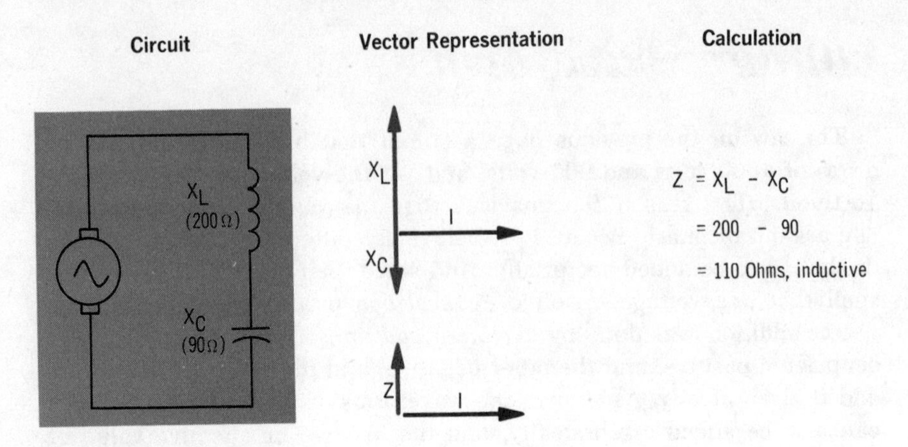

impedance

You will recall from series RL and RC circuits that it was convenient to consider resistance, inductive reactance, and capacitive reactance as vector quantities because of the phase relationships between the voltage drops they cause and the circuit current. The direction of these vector quantities was then the same as the direction of their respective voltage drops. And as a result, the impedance vector had the same direction as the applied voltage. These same techniques are used to determine the impedance of series LC circuits.

The inductive reactance, like the voltage across the inductance (E_L), leads the circuit current (which is the 0-degree reference) by 90 degrees, and the capacitive reactance, like the voltage across the capacitance (E_C), lags the current by 90 degrees. Since the current is the same in both, the inductive reactance is 180 degrees out of phase with the capacitive reactance. The impedance is then the vector sum of the two reactances. The reactances are 180 degrees apart, so their vector sum is found by subtracting the smaller one from the larger. If the inductive reactance is the larger, the equation for impedance is $Z = X_L - X_C$. And if the capacitive reactance is the larger, the equation is $Z = X_C - X_L$.

Unlike RL and RC circuits, in which the impedance was a combination of resistance and reactance, the impedance in an LC circuit is either purely inductive or purely capacitive. It is purely inductive if X_L is the larger reactance, and purely capacitive if X_C is the larger. Usually, the type of the impedance is specified directly after the impedance value. For example, an impedance of 50 ohms, capacitive; or 10 ohms, inductive. If the impedance is capacitive, the current is purely capacitive; and if the impedance is inductive, the current is purely inductive.

current

The same current flows through both the inductance and capacitance in a series LC circuit. If the *inductive* reactance (X_L) is the *greater* of the two circuit reactances, the current is purely inductive, and *lags* the applied voltage by 90 degrees. And if the *capacitive* reactance (X_C) is the *larger* reactance, the current is purely capacitive, and *leads* the applied voltage by 90 degrees. As far as the current and applied voltage are concerned, therefore, a series LC circuit is either a purely inductive or purely capacitive circuit: inductive if the impedance, which is the net reactance, is inductive, and capacitive if the impedance is capacitive. The magnitude of the current is related to the applied voltage and impedance by Ohm's Law for a-c circuits. So if the applied voltage and impedance are known, the current can be calculated from the equation:

$$I = E_{APP}/Z$$

The thought may have occurred to you: "What happens if the two reactances are *equal?*" This is entirely possible, and as a matter of fact is very often the case. If the two reactances were equal in a purely

The magnitude of the current in series LC circuits is determined by Ohm's Law

$I = E_{APP}/Z$

The phase angle between the current and applied voltage depends on the relative values of X_L and X_C

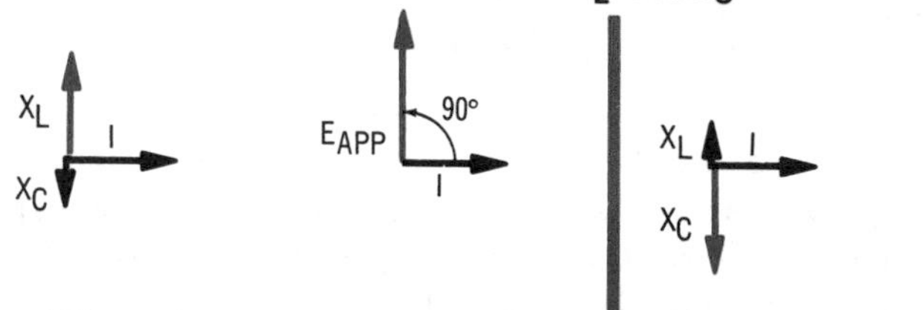

If X_L is the larger reactance, the current lags the voltage by 90°

If X_C is the larger reactance, the current leads the voltage by 90°

LC circuit, which we are considering now, the impedance would be *zero.* So with no opposition, an infinitely large current would flow. Of course this is never the case, since every circuit has some resistance. In actual circuits, when the two reactances are equal, large currents do flow, with the magnitude of the current limited only by the circuit resistance. This condition is called *resonance,* and will be covered later.

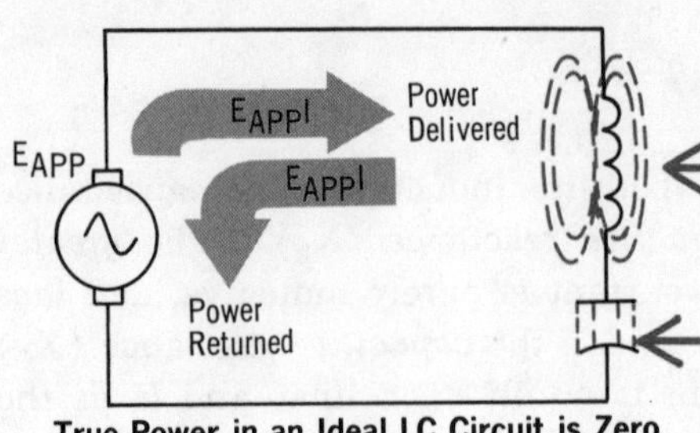

True Power in an Ideal LC Circuit is Zero

The power delivered to the inductance is stored in the magnetic field when the field is expanding, and returned to the source when the field collapses

The power delivered to the capacitance is stored in the electrostatic field when the capacitor is charging, and returned to the source when the capacitor discharges

None of the power delivered to the circuit by the source is consumed. It is all returned to the source. The true power, which is the power consumed, is thus zero

power

In a series LC circuit, power is delivered by the source to both the inductance and capacitance. The energy represented by the power delivered to the inductance is *stored* in the *magnetic field* around the inductance, and that delivered to the capacitance is *stored* in the *electrostatic field* between the capacitor plates. Since theoretically there is no resistance in the circuit, all of the stored energy is *returned* to the source each time the magnetic field around the inductance collapses and the capacitor discharges. There is thus a constant interchange of power, or energy, between the source and the circuit, but no power *consumption*. Inasmuch as true power is the power actually consumed, or dissipated, in a circuit, the *true power* in a pure LC circuit is *zero*. The apparent power, which is the total power delivered by the source, is the sum of the powers delivered to the inductance and capacitance. It is equal to the applied voltage times the circuit current, or:

$$P_{APPARENT} = E_{APP}I$$

The circuit power factor is equal to the cosine of the phase angle between the applied voltage and the current, as it is in any a-c circuit. Since the phase angle in a series LC circuit is always 90 degrees, and the cosine of 90 degrees is zero, the power factor in an LC circuit is zero. True power in an a-c circuit equals the apparent power times the power factor; this is another way of showing that the true power in a series LC circuit is always zero.

$$\begin{aligned} P_{TRUE} &= E_{APP}I \cos \theta \\ &= E_{APP}I \times 0 = 0 \end{aligned}$$

Actually, all circuits contain some resistance, of course, so the power factor is never zero because of the power dissipated by the resistance. When the resistance is very low, though, for all practical purposes the power factor can be considered zero.

effect of frequency

In series RL and RC circuits, a definite relationship exists between the *impedance,* and the *frequency* of the applied voltage. *Increasing* the frequency of an *RL* circuit, or *decreasing* the frequency of an *RC* circuit, results in an *increase* in *impedance.* Similarly, *decreasing* the frequency in an *RL* circuit, or *increasing* the frequency in an *RC* circuit, causes a *decrease* in *impedance.*

In series LC circuits, however, no similar clear-cut relationship exists between the impedance, and the frequency. The impedance is controlled by the frequency, but an increase or decrease in the frequency depends on the *relative values* of the inductive and capacitive reactance. For example, in one circuit, an increase in frequency may

The effect of frequency changes on series LC circuits depends on the relative valves of X_L and X_C

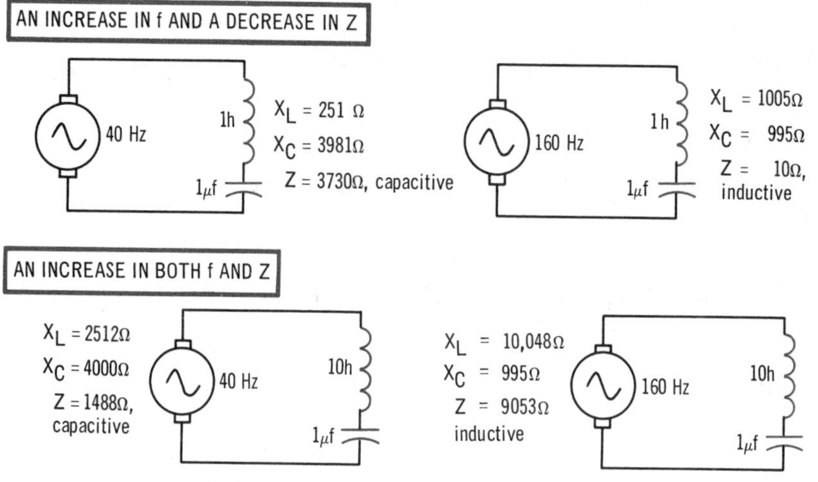

Although the impedance is controlled by the frequency, no relationship can be specified. As shown, the identical increase in frequency can cause an increase in impedance in one circuit, and a decrease in another

cause an increase in impedance, while in another circuit, the same increase in frequency could cause a decrease in impedance. In the same way, a decrease in capacitance can cause an increase in impedance in one circuit and a decrease in another. The situation can even occur where a certain increase, say in inductance, will cause a decrease in impedance, but a further increase in the inductance will cause the impedance to increase. It is, therefore, usually best to actually calculate the new impedance whenever a change occurs in the frequency of a series LC circuit.

solved problems

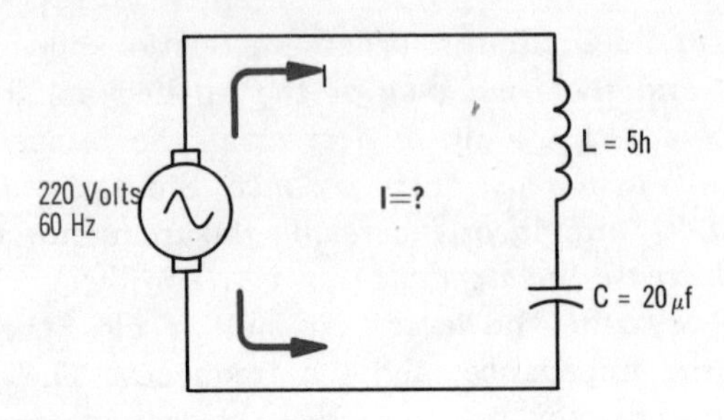

Problem 18. What is the current in the circuit?

According to Ohm's Law for a-c circuits, the equation for current is $I = E_{APP}/Z$. The voltage is given, but the impedance is not, and must therefore be calculated. Before you can calculate the impedance, though, the two reactances, X_L and X_C, must be determined. Thus, the sequence for solving this problem is to first calculate X_L and X_C, then Z, and finally I.

Calculating X_L and X_C:

$$X_L = 2\pi fL = 6.28 \times 60 \times 5 = 1884 \text{ ohms}$$

$$X_C = \frac{1}{2\pi fC} = \frac{1}{6.28 \times 60 \times 0.00002} = 133 \text{ ohms}$$

Calculating Z: X_L is larger than X_C, so the equation for Z is

$$Z = X_L - X_C = 1884 - 133 = 1751 \text{ ohms, inductive}$$

Calculating I:

$$I = E_{APP}/Z = 220 \text{ volts}/1751 \text{ ohms} = 0.125 \text{ ampere}$$

Problem 19. What is the phase angle between the applied voltage and current in the above circuit?

The phase angle in all purely LC circuits is 90 degrees; thus, in this circuit, the angle must be 90 degrees. And since the impedance was inductive, the applied voltage leads the current.

Problem 20. What are the voltage drops across the reactances?

Each voltage drop is independent, and depends only on the circuit current and the value of reactance. So,

$$E_L = IX_L = 0.125 \times 1884 = 236 \text{ volts}$$

$$E_C = IX_C = 0.125 \times 133 = 17 \text{ volts}$$

Notice that the voltage drop across the inductance (E_L) is greater than the applied voltage of 220 volts.

solved problems (cont.)

Problem 21. Which of the following circuits are inductive and which are capacitive?

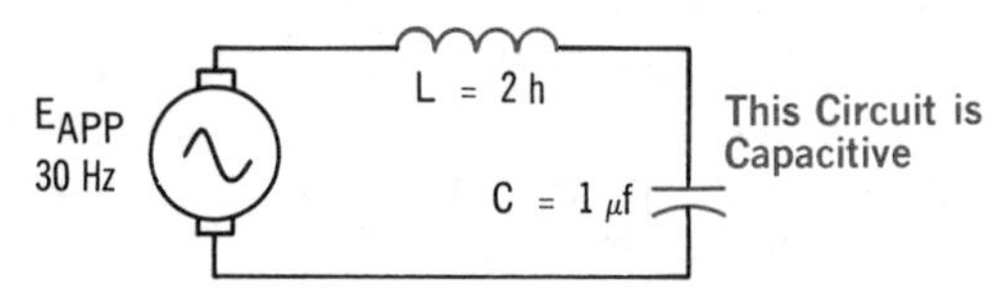

$$X_L = 2\pi fL = 6.28 \times 30 \times 2 = 377 \text{ ohms}$$

$$X_C = \frac{1}{2\pi fC} = \frac{1}{6.28 \times 30 \times 0.000001} = 5300 \text{ ohms}$$

$$Z = X_C - X_L = 5300 - 377 = 4923 \text{ ohms, capacitive}$$

EAPP 60 Hz, L = 5 h, C = 0.01 μf — This Circuit is Capacitive

$$X_L = 2\pi fL = 6.28 \times 60 \times 5 = 1884 \text{ ohms}$$

$$X_C = \frac{1}{2\pi fC} = \frac{1}{6.28 \times 60 \times 0.00000001} = 265{,}000 \text{ ohms}$$

$$Z = X_C - X_L = 265{,}000 - 1884 \cong 263\text{K, capacitive}$$

EAPP 1000 Hz, L = 5 mh, C = 20 μf — This Circuit is Inductive

$$X_L = 2\pi fL = 6.28 \times 1000 \times 0.005 = 31 \text{ ohms}$$

$$X_C = \frac{1}{2\pi fC} = \frac{1}{6.28 \times 1000 \times 0.000020} = 8 \text{ ohms}$$

$$Z = X_L - X_C = 31 - 8 = 23 \text{ ohms, inductive}$$

EAPP 200 Hz, L = 200 mh, C = 0.1 μf — This Circuit is Capacitive

$$X_L = 2\pi fL = 6.28 \times 200 \times 0.2 = 251 \text{ ohms}$$

$$X_C = \frac{1}{2\pi fC} = \frac{1}{6.28 \times 200 \times 0.0000001} = 7900 \text{ ohms}$$

$$Z = X_C - X_L = 7900 - 251 = 7649 \text{ ohms, capacitive}$$

summary

☐ A series LC circuit consists of an inductance and capacitance in series, with no resistance in the circuit. ☐ Since the current is the same throughout the circuit, it is used as the phase reference. ☐ The voltage across the inductor in a series LC circuit leads the current through it by 90 degrees. It is equal to: $E_L = IX_L$. ☐ The voltage across the capacitor lags the current through it by 90 degrees. It is equal to: $E_C = IX_C$. ☐ Because the voltages are 180 degrees out of phase with each other, the total, or applied, voltage is equal to the arithmetic difference of the two. ☐ The phase angle is determined by the larger of the two voltage drops.

☐ In a series LC circuit, either E_L or E_C, or in some cases both, are greater than the applied voltage. ☐ The impedance of a series LC circuit is either purely inductive or purely capacitive, depending upon the magnitudes of the reactances. Usually, the type of impedance is specified after the impedance value. ☐ Ohm's Law for a-c circuits applies in series LC circuits: $I = E_{APP}/Z$.

☐ In a pure series LC circuit, the true power is zero. ☐ The power factor is also zero. ☐ Because both X_L and X_C are dependent on frequency, there is no definite relationship between impedance and the frequency. The impedance depends on the relative values of the inductive and capacitive reactances in the circuit.

review questions

1. In a series LC circuit, can either the voltage across L and/or C ever be greater than the applied voltage?

For Questions 2 to 10, consider a series LC circuit with an applied voltage of 100 volts; a capacitor with a voltage drop of 140 volts; and an inductor with an inductive reactance of 20 ohms.

2. What is the current in the circuit?
3. What is the impedance of the circuit?
4. What is the apparent power of the circuit?
5. What is the true power of the circuit?
6. What is the phase angle of the circuit?
7. What is the power factor of the circuit?
8. What is the capacitive reactance?
9. Answer Questions 2 to 8, where the frequency of the applied voltage is doubled. (*Hint:* The voltage across the capacitor is no longer 140 volts.)
10. Answer Questions 2 to 8, where the frequency of the applied voltage is halved.

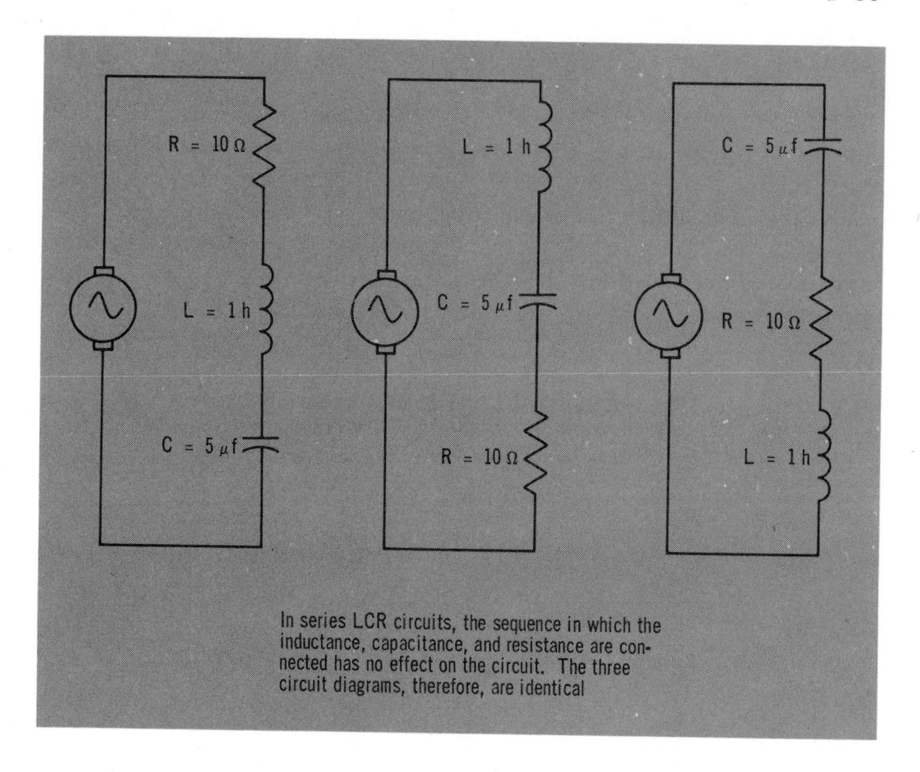

In series LCR circuits, the sequence in which the inductance, capacitance, and resistance are connected has no effect on the circuit. The three circuit diagrams, therefore, are identical

series LCR circuits

Any practical series LC circuit contains some resistance. When the resistance is very small compared to the circuit reactances, it has almost no effect on the circuit and can be considered as being zero, which is what was done on the preceding pages. When the resistance is *appreciable*, though, it has a significant effect on the circuit operation, and therefore must be considered in any circuit analysis. It makes no difference whether the resistance is the result of the circuit wiring or coil windings, or whether it is in the form of a resistor connected into the circuit. As long as it is appreciable, it affects the circuit operation and so must be considered. As a general rule, if the total reactance of the circuit is *not* 10 times or more greater than the resistance, the resistance *will* have an effect.

Circuits in which the inductance, capacitance, and resistance are all connected in series are called series LCR ciruits. You will see that the fundamental properties of series LCR circuits, and the methods used to solve them, are similar to those that you have learned for series LC circuits. The differences are caused by the effects of the resistance.

voltage

Since there are three elements in a series LCR circuit, there are three voltage drops around the circuit: one across the inductance, one across the capacitance, and the other across the resistance. The same current flows through each circuit element, so the phase relationships between the voltage drops are the same as they are in series LC, RL, and RC circuits. The voltage drops across the inductance and capacitance

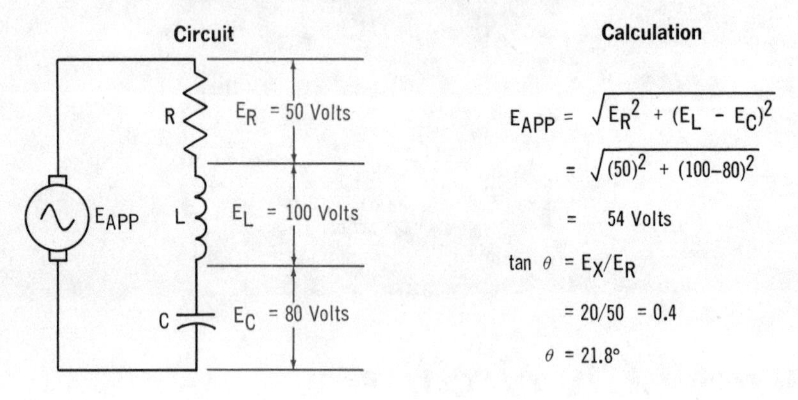

are 180 degrees out of phase, with the inductive voltage drop (E_L) leading the resistive voltage drop (E_R) by 90 degrees, and the capacitive voltage drop (E_C) lagging the resistive voltage drop (E_R) by 90 degrees.

The vector sum of the three voltage drops is equal to the applied voltage. However, to calculate this vector sum, a *combination* of the methods you have learned for LC, RL, and RC circuits must be used. You first have to calculate the combined voltage drop of the two reactances. This value is designated E_X, and is found, as in pure LC circuits, by subtracting the smaller reactive voltage drop from the larger. The result of this calculation is the net reactive voltage drop, and is either inductive or capacitive, depending on which of the individual voltage drops was larger. As an equation, the net reactive voltage drop can be written:

$$E_X = E_L - E_C$$

if E_L is larger than E_C; or

$$E_X = E_C - E_L$$

if E_C is larger than E_L.

voltage (cont.)

Once the net reactive voltage drop is known, it is added vectorially to the voltage drop across the resistance, using the Pythagorean Theorem. The equation for this vector addition is

$$E_{APP} = \sqrt{E_R^2 + E_X^2}$$

The vector addition of all three voltage drops can be put into one equation by substituting in the above equation the values of E_X given on the previous page. Thus,

$$E_{APP} = \sqrt{E_R^2 + (E_L - E_C)^2}$$

if E_L is larger than E_C; or

$$E_{APP} = \sqrt{E_R^2 + (E_C - E_L)^2}$$

if E_C is larger than E_L.

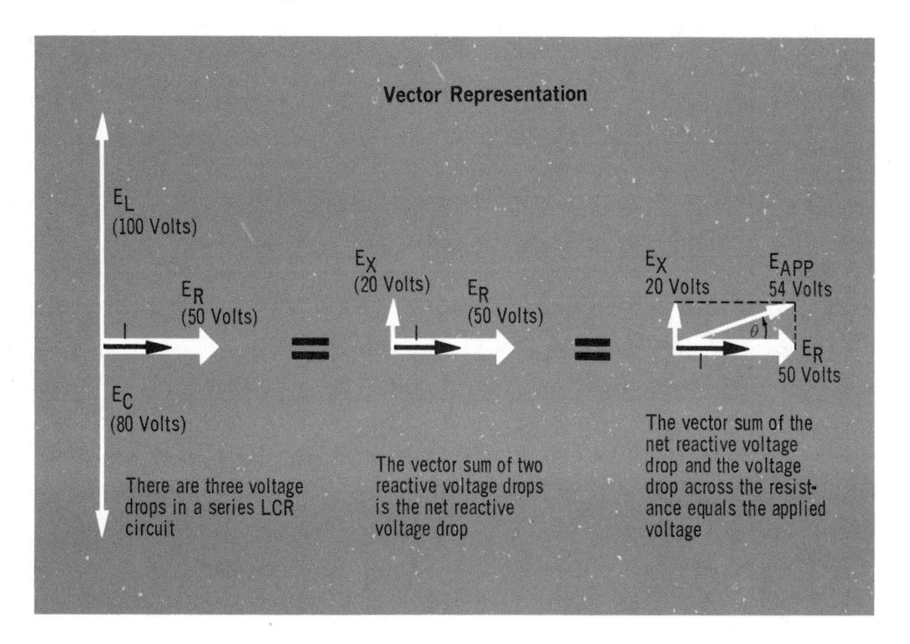

As you can see from the vector diagrams shown, the angle between the applied voltage (E_{APP}) and the voltage across the resistance (E_R) is the same as the phase angle between the applied voltage and the circuit current. The reason for this is that E_R and I are in phase. The value of the phase angle can be found from:

$$\tan \theta = E_X/E_R$$

voltage waveforms

The voltage waveforms in a series LCR circuit are a combination of those in series RL, RC, and LC circuits. The applied voltage waveform is the sum of the instantaneous values of three voltage waveforms, all 90 degrees out of phase, rather than of two voltages less than 90 degrees out of phase, as in RL and RC circuits, or of two voltages 180 degrees out of phase, as in LC circuits. Because of their different phase relationship, the vector addition of the three voltage drops in an LCR circuit has to be done in two steps; first the two reactive voltage drops, and then their resultant and the resistive voltage drop. When they are represented as waveforms, though, the three voltage drops can be shown to add simultaneously to produce the applied voltage waveform.

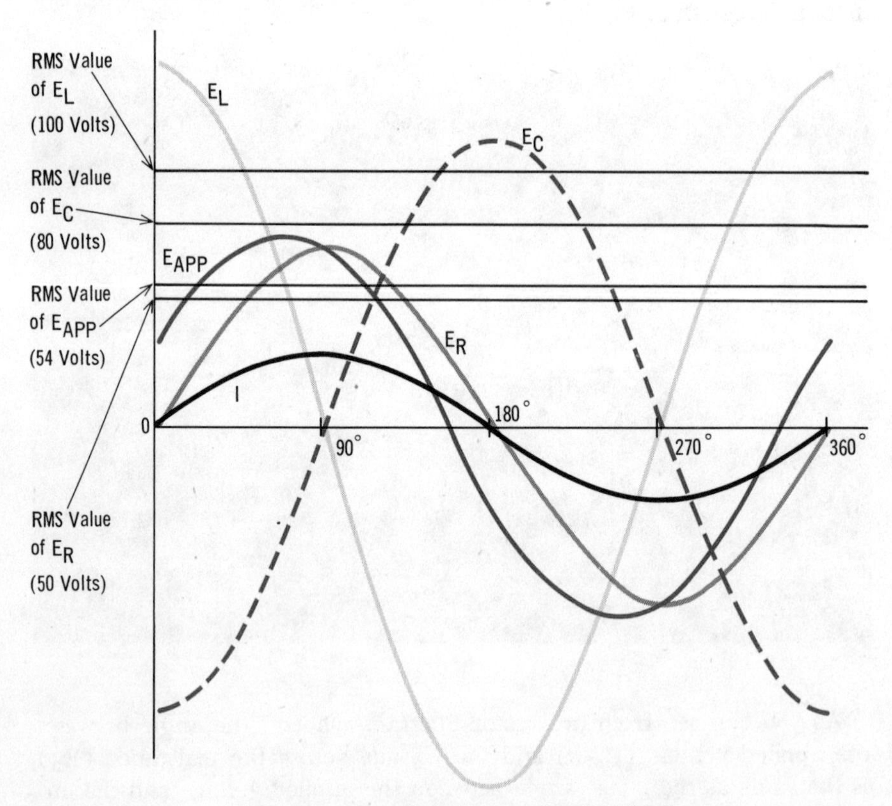

Every point on the applied voltage waveform (E_{APP}) is the algebraic sum of the instantaneous values of the E_L, E_C, and E_R waveforms

The impedance of a series LCR circuit is the VECTOR SUM of the inductive reactance, the capacitive reactance, and the resistance

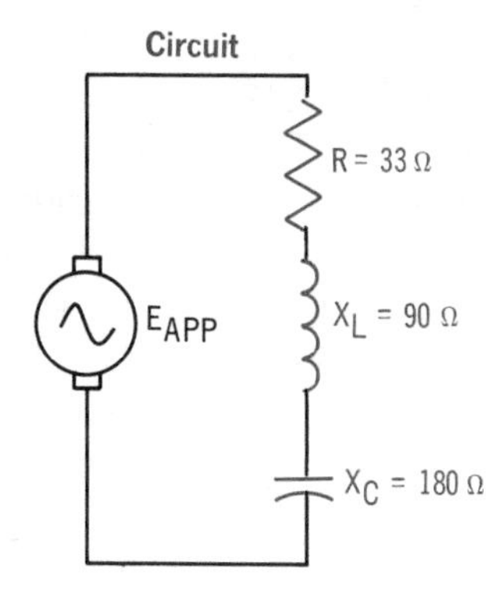

Calculation

$$Z = \sqrt{R^2 + (X_C - X_L)^2}$$
$$= \sqrt{(33)^2 + (180 - 90)^2}$$
$$= 96 \text{ Ohms, capacitive}$$

$$\tan \theta = X/R$$
$$= (180-90)/33 = 2.72$$
$$\theta = 69.8°$$

impedance

The impedance of a series LCR circuit is the vector sum of the inductive reactance, the capacitive reactance, and the resistance. This vector addition is the same type as you have just learned for adding the voltage drops around a series LCR circuit. The two reactances are 180 degrees out of phase, so the net reactance, designated X, is found by subtracting the smaller reactance from the larger. Therefore,

$$X = X_L - X_C$$

if X_L is larger than X_C; or

$$X = X_C - X_L$$

if X_C is larger than X_L.

The impedance is then the vector sum of the *net reactance* and the *resistance,* and is calculated by the Pythagorean Theorem:

$$Z = \sqrt{R^2 + X^2}$$

If the equations for the net reactance and impedance are combined, the impedance can be calculated from a single equation, which is

$$Z = \sqrt{R^2 + (X_L - X_C)^2}$$

if X_L is larger than X_C; or

$$Z = \sqrt{R^2 + (X_C - X_L)^2}$$

if X_C is larger than X_L.

impedance (cont.)

When X_L is greater than X_C, the net reactance is inductive, and the circuit acts essentially as an *RL circuit.* This means that the impedance, which is the vector sum of the net reactance and resistance, will have an angle between 0 and 90 degrees. Similarly, when X_C is greater than X_L, the net reactance is capacitive, and the circuit acts as an *RC circuit.* The impedance, therefore, has an angle somewhere between 0 and 90 degrees. In both cases, the value of the impedance angle depends on the relative values of the net reactance (X) and the resistance (R). The angle can be found by:

$$\tan \theta = X/R$$

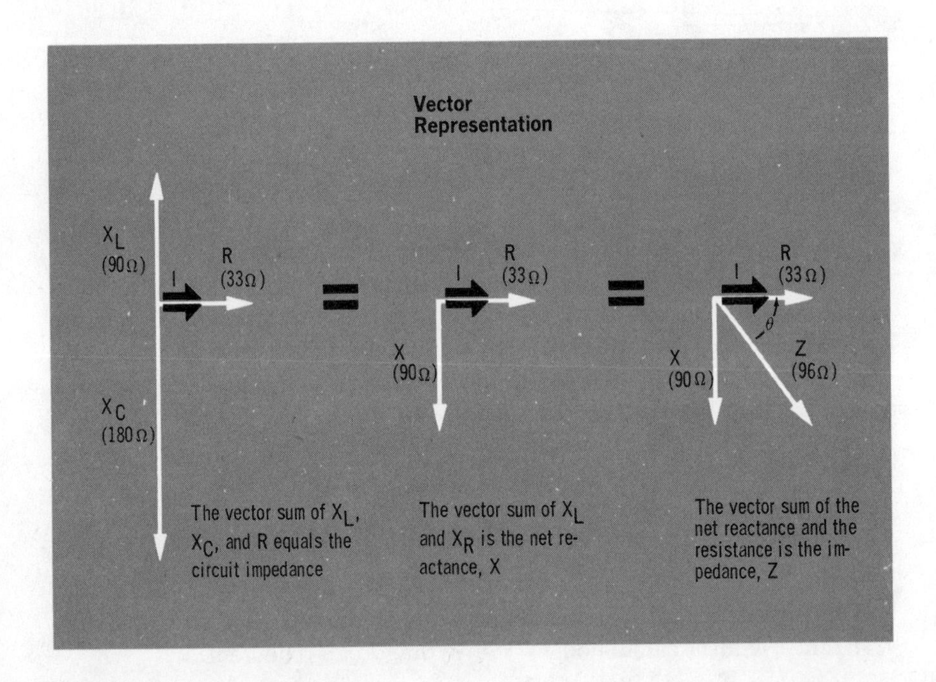

A point you should notice about the impedance of a series LCR circuit is that its value depends on the resistance and the *relative* values of X_L and X_C. High reactances do not necessarily mean a high impedance. A circuit can have very large reactances, but if their difference, or X, is small, the impedance will be low for a given value of resistance. And if R is greater than X, the impedance will be more resistive. The 10-to-1 rule applies to X and R, as it does to X_L or X_C and R in an RL or RC circuit.

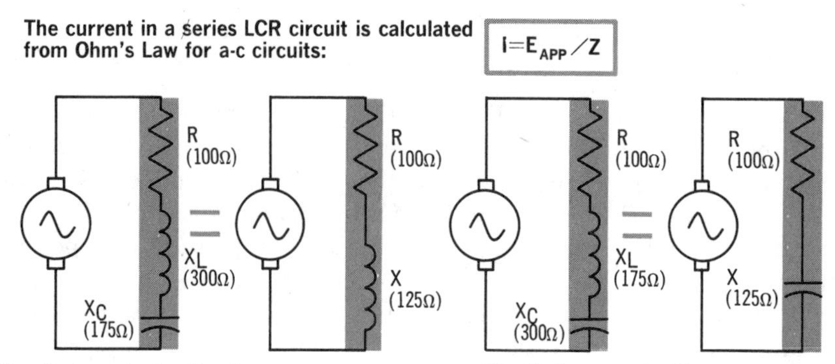

When X_L is greater than X_C, the current acts as an RL circuit, so the current lags the voltage

When X_C is greater than X_L, the circuit acts as an RC circuit, so the current leads the voltage

current

The same current flows in every part of a series LCR circuit. If the impedance and the applied voltage are known, the magnitude of the current can be calculated by Ohm's Law for a-c circuits:

$$I = E_{APP}/Z$$

The current always leads the voltage across the capacitance by 90 degrees, lags the voltage across the inductance by 90 degrees, and is in phase with the voltage across the resistance. The phase relationship between the current and the applied voltage, however, depends on the circuit *impedance*. If the impedance is *inductive* (X_L greater than X_C), the current is inductive, and *lags* the applied voltage by some phase angle less than 90 degrees. And if the impedance is *capacitive* (X_C greater than X_L), the current is capacitive, and *leads* the applied voltage by some phase angle also less than 90 degrees. The angle of the lead or lag is determined by the relative values of the net reactance and the resistance according to the equation:

$$\tan \theta = X/R$$

The greater the value of X, or the smaller the value of R, the larger is the phase angle, and the more reactive (or less resistive) is the current. Similarly, the smaller the value of X, or the larger the value of R, the more resistive (or less reactive) is the current. If either R or X is 10 or more times greater than the other, the circuit will act essentially as though it was purely resistive or reactive, as the case may be.

Other useful equations for calculating the phase angle can be derived from the vector diagrams for impedance and applied voltage. Two of these equations are

$$\cos \theta = R/Z \qquad \tan \theta = E_X/E_R$$

power

In a *pure* LC circuit, you will recall that the true power is zero, since all of the power delivered by the source is returned to it. In a series LCR circuit, the power delivered to the inductance and capacitance is also returned to the source, but in addition, power is *dissipated* by the *resistance* in the form of I^2R heating. This power represents true power, since by definition, true power is the power dissipated, or "used-up," in the circuit. The amount of true power depends on the value of the resistance and the current flow. As was pointed out previously, the impedance of a series LCR circuit, and therefore the circuit current, is determined in large part by how *close* the values of X_L and X_C are. The closer they are, the lower is the impedance, the greater is the circuit current, and the larger is the true power for a given resistance. You can see, then, that anything that affects the relative values of X_L and X_C will also affect the power dissipated in the circuit.

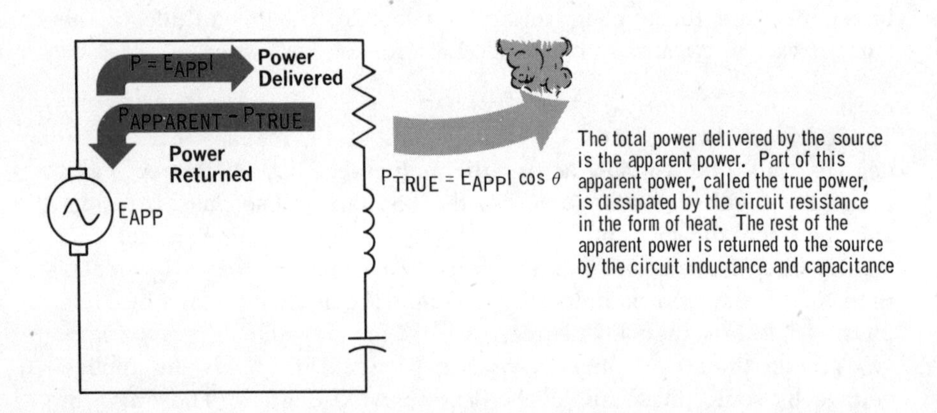

The total power delivered by the source is the apparent power. Part of this apparent power, called the true power, is dissipated by the circuit resistance in the form of heat. The rest of the apparent power is returned to the source by the circuit inductance and capacitance

The value of the true power in a series LCR circuit can be calculated from the standard equation for power in a-c circuits:

$$P_{TRUE} = E_{APP}I \cos \theta$$

The equation can also be written as $P_{TRUE} = I^2Z \cos \theta$; and since from a vector diagram for impedance, $R = Z \cos \theta$, true power can also be expressed as:

$$P_{TRUE} = I^2R$$

The apparent power, which is the total power *delivered* by the source, is simply equal to the applied voltage times the circuit current, or

$$P_{APPARENT} = E_{APP}I$$

effect of frequency

You will recall that in series LC circuits, although frequency affects the characteristics of the circuit, there is no clear-cut relationship between an increase or decrease in frequency and a corresponding increase or decrease in circuit impedance or current. The reason for this is that when frequency is increased, X_L also increases, but X_C decreases; and when frequency is decreased, X_C increases while X_L decreases. The impedance, on the other hand, varies as the *difference* between X_L and X_C. So, whether the impedance increases or decreases depends on what the *relative* values of X_L and X_C were before the frequency change took place.

As you will learn later, for every series LCR circuit there is one frequency, called the *resonant frequency,* at which X_L and X_C are equal. The frequency is determined by the values of the inductance and capacitance, and is unaffected by the circuit resistance. Any change in frequency *away* from the resonant frequency will result in an increase in the net reactance and the impedance, and a resulting decrease in current. A change which brings the frequency *closer* to the resonant frequency will have the opposite effect. Net reactance and impedance will decrease, so circuit current will increase. This is covered later.

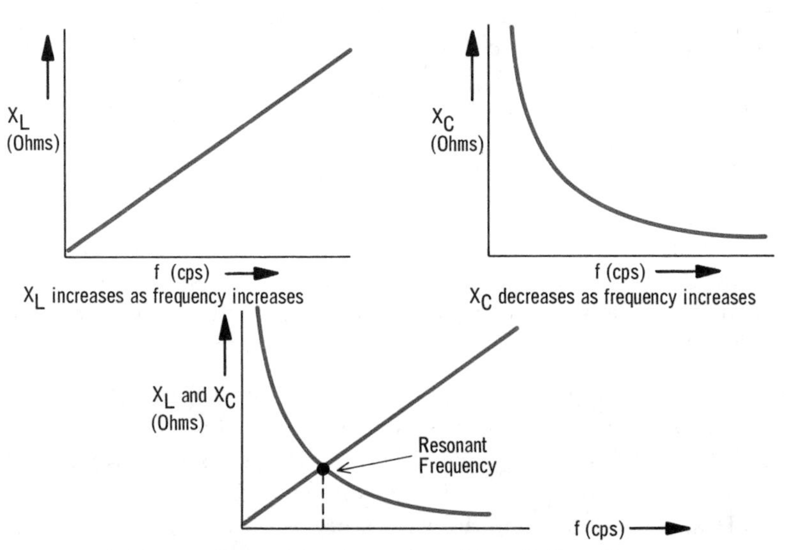

For every combination of inductance and capacitance, the X_L and X_C curves intersect at one point, where the values of X_L and X_C are equal, and which corresponds to the resonant frequency. Any change in frequency from this point results in an increase in impedance. Conversely, if the frequency comes closer to this point, impedance decreases

solved problems

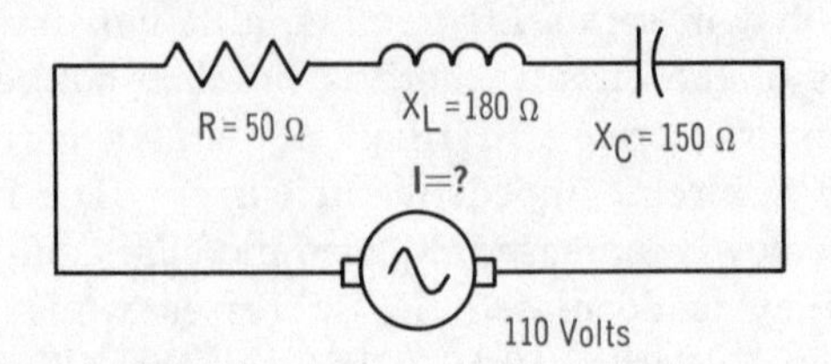

Problem 22. What is the current in the circuit?

The current is calculated from the equation $I = E_{APP}/Z$. The applied voltage is given, but the impedance has to be determined before the current can be calculated.

Calculating Z: X_L is larger than X_C, so the equation for Z is

$$Z = \sqrt{R^2 + (X_L - X_C)^2} = \sqrt{(50)^2 + (180 - 150)^2} = 58 \text{ ohms, inductive}$$

Calculating I:

$$I = E_{APP}/Z = 110 \text{ volts}/58 \text{ ohms} = 1.9 \text{ amperes}$$

Problem 23. What is the relationship between the current and the applied voltage?

With the equation that is known, the phase angle, θ, between the current and the applied voltage can be calculated with either of the equations: $\tan\theta = X/R$, or $\cos\theta = R/Z$.

$$\tan\theta = \frac{X}{R} = \frac{X_L - X_C}{R} = \frac{30}{50} = 0.6 \qquad \theta = 31°$$

$$\cos\theta = R/Z = 50/58 = 0.862 \qquad \theta = 31°$$

The impedance of the circuit is inductive, since X_L is greater than X_C, so the current lags the applied voltage, as it does in all inductive circuits. A complete description of the phase relationship between the current and the applied voltage, therefore, is that the current lags the voltage by 31 degrees.

Problem 24. How much power is consumed in the circuit?

Consumed power is true power, and you know that the true power can be calculated from the equation $P_{TRUE} = E_{APP}I\cos\theta$. However, you also know that another form of the equation for the true power is $P_{TRUE} = I^2R$. Since you know the values of both I and R, you can use this equation. Very often, by using this equation you will avoid the necessity of finding the value of θ in trigonometric tables.

$$P_{TRUE} = I^2R = (1.9)^2 \times 50 = 181 \text{ watts}$$

solved problems (cont.)

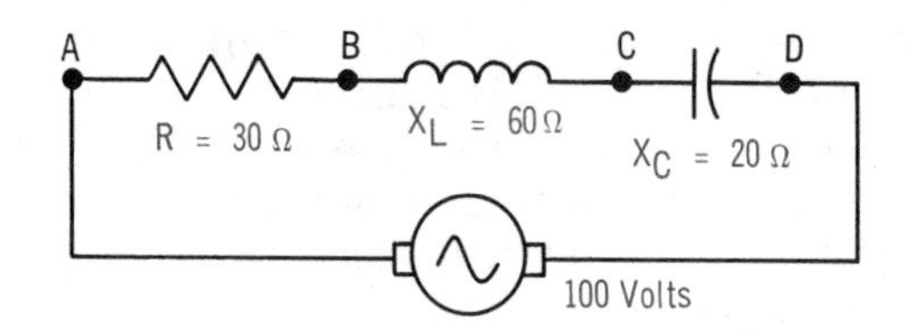

Problem 25. What are the voltage drops between AB, BC, CD, BD, and AC?

Before the voltage drops can be found, the circuit current has to be calculated; but before this can be done, the impedance must be determined. X_L is larger than X_C, so the impedance is calculated from the equation:

$$Z = \sqrt{R^2 + (X_L - X_C)^2} = \sqrt{(30)^2 + (60 - 20)^2}$$
$$= \sqrt{2500} = 50 \text{ ohms, inductive}$$

The current can now be found by using Ohm's Law:

$$I = E_{APP}/Z = 100 \text{ volts}/50 \text{ ohms} = 2 \text{ amperes}$$

Knowing the current, you can now calculate the voltage drops:

Calculating Voltage Drop AB: The voltage drop across the resistance is equal to the current times the resistance.

$$E_R = IR = 2 \text{ amperes} \times 30 \text{ ohms} = 60 \text{ volts}$$

Calculating Voltage Drop BC: The voltage drop across the inductance is equal to the current times the inductive reactance.

$$E_L = IX_L = 2 \text{ amperes} \times 60 \text{ ohms} = 120 \text{ volts}$$

Calculating Voltage Drop CD: The voltage drop across the capacitance is equal to the current times the capacitive reactance.

$$E_C = IX_C = 2 \text{ amperes} \times 20 \text{ ohms} = 40 \text{ volts}$$

Calculating Voltage Drop BD: Voltage drop BD is the net voltage, E_X, across the two reactances. It is equal to the difference between the two individual voltages E_L and E_C.

$$E_X = E_L - E_C = 120 - 40 = 80 \text{ volts}$$

Calculating Voltage Drop AC: Voltage drop AC is the vector sum of the voltage drops across the resistance and the inductance. These two voltages are 90 degrees out of phase, so they can be added vectorially by the Pythagorean Theorem:

$$E_{AC} = \sqrt{E_R^2 + E_L^2} = \sqrt{(60)^2 + (120)^2} = 134 \text{ volts}$$

summary

☐ A series LCR circuit consists of an inductor, a resistor, and a capacitor in series. ☐ The net reactance of the circuit determines whether the circuit behaves like an RL or RC circuit. ☐ The total reactive voltage of a series LCR circuit is given by: $E_X = E_L - E_C$, for E_L greater than E_C; and $E_X = E_C - E_L$, for E_C greater than E_L. ☐ The applied voltage is equal to the vector sum of the resistor voltage and the total reactive voltage: $E_{APP} = \sqrt{E_R + E_X^2}$. ☐ The phase angle is given by $\tan\theta = E_X/E_R$; or $\tan\theta = X/R$; or $\cos\theta = R/Z$.

☐ The impedance of a series LCR circuit is the vector sum of the resistance and net reactance: $Z = \sqrt{R^2 + X^2}$, where X is the net reactance, and is equal to $X_L - X_C$, or $X_C - X_L$. ☐ The current is equal to: $I = E_{APP}/Z$. It leads or lags the applied voltage depending on the net reactance of the circuit.

☐ Unlike a pure LC circuit, the LCR circuit does dissipate power in the resistor: $P_{TRUE} = E_{APP}I\cos\theta = I^2Z\cos\theta = I^2R$. ☐ The apparent power is equal to: $P_{APPARENT} = E_{APP}I$. ☐ The true power can also be found by $P_{TRUE} = P_{APPARENT} \times$ power factor. ☐ The frequency at which the net reactance is zero, or where X_L equals X_C, is known as the resonant frequency.

review questions

1. Can the voltage across the inductor or the capacitor in a series LCR circuit ever be greater than the applied voltage?
2. Can the voltage across the resistor in a series LCR circuit ever be greater than the applied voltage?

For Questions 3 to 8, consider a series LCR circuit with an applied voltage of 200 volts, an impedance of 100 ohms, an inductive reactance of 50 ohms, and a capacitive reactance of 130 ohms.

3. What is the value of the resistance?
4. What is the phase angle of the circuit?
5. What is the current in the circuit?
6. What is the apparent power? The true power?
7. What would be the impedance if the frequency were doubled?
8. What would be the impedance if the frequency were halved?
9. The applied voltage across a series LCR circuit is 200 volts, the voltage across the resistor is 160 volts, and the voltage across the inductor is 300 volts. What values can exist across the capacitor?
10. Answer Question 9, where the voltage across the inductor is 100 volts. (*Hint:* There is only one solution.)

series resonance

Resonance was briefly described as a condition that occurs when the inductive reactance and capacitive reactance of a series LCR circuit are *equal.* When this happens, the two reactances, in effect, *cancel* each other, and the impedance of the circuit is equal to the resistance. Current, therefore, is opposed only by the resistance, and if the resistance is relatively low, *very large currents* can flow. Remember, though, that the two reactances cancel each other only as far as their opposition to current is concerned. They are still present in the circuit, and because of the large current that flows when they are equal, extremely high voltages drops exist across them.

Series Resonance Occurs When:
X_L EQUALS X_C

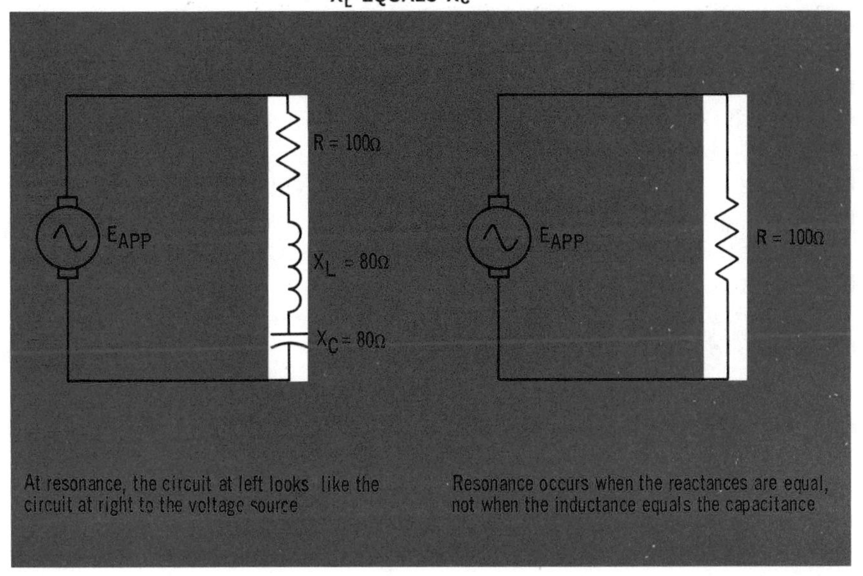

At resonance, the circuit at left looks like the circuit at right to the voltage source

Resonance occurs when the reactances are equal, not when the inductance equals the capacitance

The two identifying characteristics of resonance in a series LCR circuit are *low impedance* and *large current.* Actually, for any given circuit, the impedance is at its minimum and the current is at its maximum at resonance.

The type of resonance that is being described is really called *series resonance.* This is to distinguish it from another type of resonance called *parallel resonance* that occurs in parallel LCR circuits. Parallel resonance is covered later.

factors that determine resonance

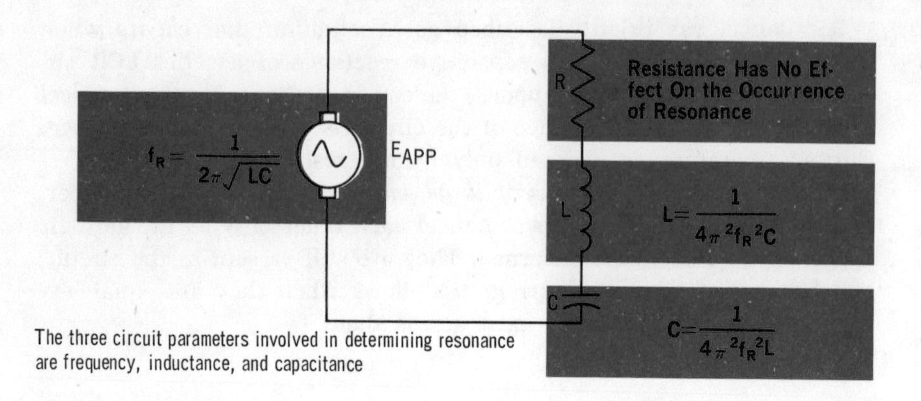

The three circuit parameters involved in determining resonance are frequency, inductance, and capacitance

Since resonance occurs when X_L and X_C are equal, resonance is affected by the *frequency* of the applied voltage, the *inductance*, and the *capacitance*. Frequency affects both X_L and X_C, while inductance affects only X_L, and capacitance only X_C.

For any given combination of an inductance and a capacitance, there is one frequency at which *X_L will equal X_C*. This frequency is called the *resonant frequency* for that particular combination. For example, a combination of a 1-henry inductance and a 4-microfarad capacitance has a resonant frequency of 80 Hz.

The values of X_L and X_C in a circuit are dependent on the frequency according to the equations:

$$X_L = 2\pi fL \qquad \text{and} \qquad X_C = \frac{1}{2\pi fC}$$

Since at resonance, X_L equals X_C, the right-hand sides of these two equations must also be equal at resonance. So,

$$2\pi fL = \frac{1}{2\pi fC}$$

When this equations is solved for f, the result is

$$f = \frac{1}{2\pi\sqrt{LC}}$$

With this equation you can find the resonant frequency, usually designated f_R, of any combination of L and C. You can also determine the value of inductance that will resonate with a particular value of capacitance at a certain frequency, and vice versa, with these equations:

$$L = \frac{1}{4\pi^2 f_R^2 C} \qquad C = \frac{1}{4\pi^2 f_R^2 L}$$

impedance at and off resonance

In most practical applications of series resonant circuits, the values of the inductance and capacitance are set, and the frequency is the variable quantity that determines whether or not a circuit is at resonance. At the resonant frequency, X_L and X_C effectively cancel each other, and the circuit impedance equals the value of the resistance.

$$Z = \sqrt{R^2 + (X_L - X_C)^2} = \sqrt{R^2 + 0} = R$$

The circuit is, therefore, completely *resistive.*

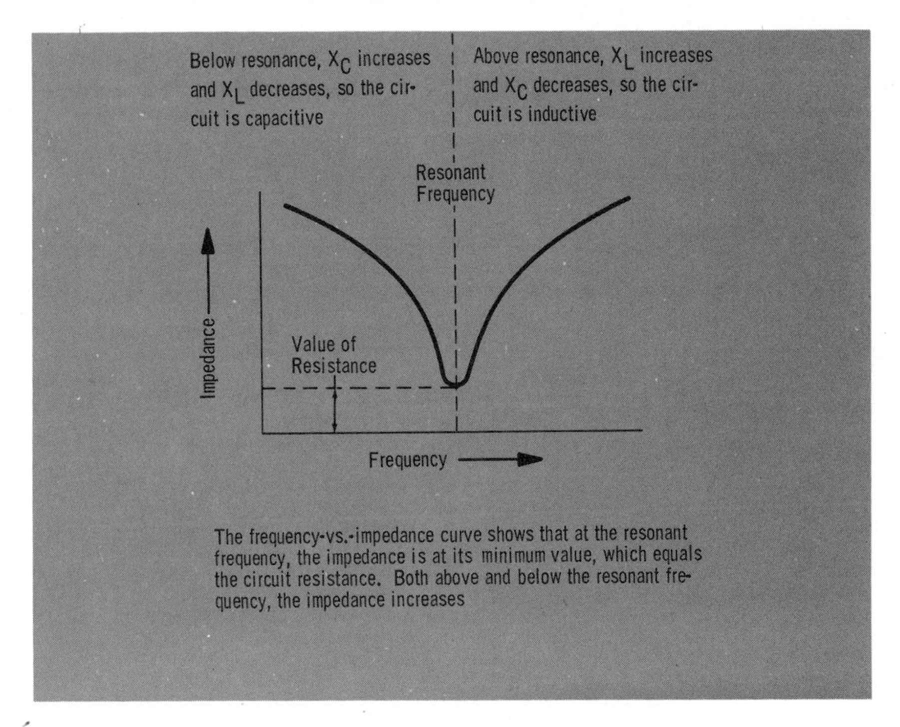

The frequency-vs.-impedance curve shows that at the resonant frequency, the impedance is at its minimum value, which equals the circuit resistance. Both above and below the resonant frequency, the impedance increases

If the frequency is varied above or below the resonant frequency, the net reactance, which is the difference between X_L and X_C, is no longer zero. The net reactance, therefore, has to be added to the resistance, and so the impedance increases. The further the frequency is varied from the resonant frequency, the greater the net reactance becomes, and the higher the impedance becomes. A characteristic of series resonant circuits is that when the frequency is *above* the resonant frequency, X_L is greater than X_C, so the circuit is *inductive;* and when the frequency is *below* the resonant frequency, X_C is greater than X_L, and the circuit is *capacitive.*

current at and off resonance

The *impedance* of a series LCR circuit is *minimum* at resonance, so the *current* must, therefore, be *maximum*. Both *above* and *below* the resonant frequency, circuit impedance increases, which means that *current decreases*. The further the frequency is from the resonant frequency, the greater is the impedance, and so the smaller the current becomes. At any frequency, the current can be calculated from Ohm's Law for a-c circuits, using the equation $I = E/Z$. Since at the resonant frequency the impedance equals the resistance, the equation for current at resonance becomes $I = E/R$.

If the frequency of an LCR circuit is varied and the values of current at the different frequencies are plotted on a graph, the result is a curve known as the *resonance curve* of the circuit.

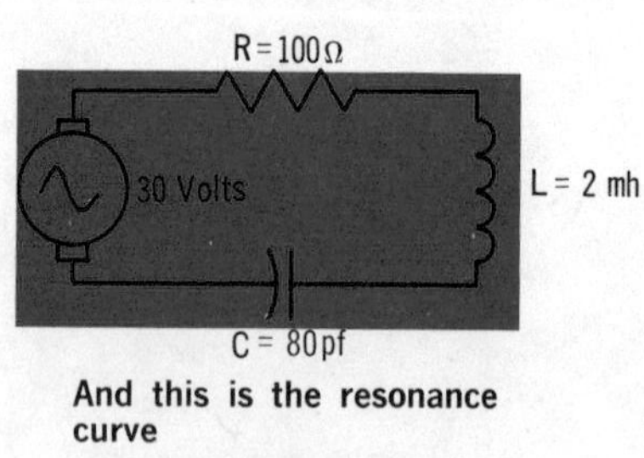

If the frequency of the applied voltage is varied from 100 to 600 kHz, these are the values of impedance and current . . .

Frequency (kHz)	Impedance (Ω)	Current (ma)
100	18,634	1.61
200	7,433	4.0
300	2,862	10.48
350	1,270	23.6
398	100	300
400	122	246
450	1,240	24.2
500	2,302	13.0
600	4,221	7.1

And this is the resonance curve

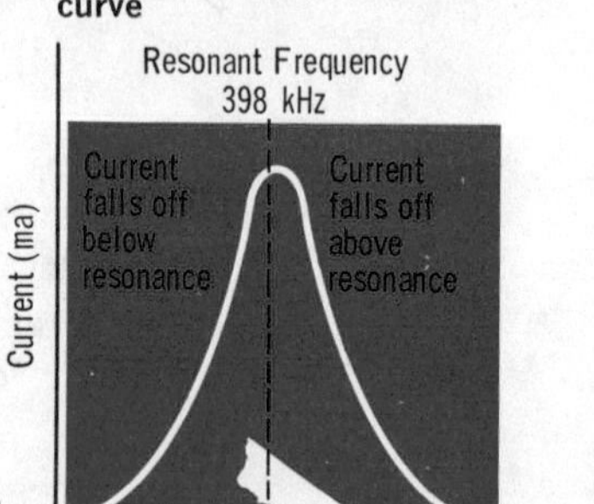

This is the characteristic shape of the resonance curve. Actual resonance curves are not always symmetrical

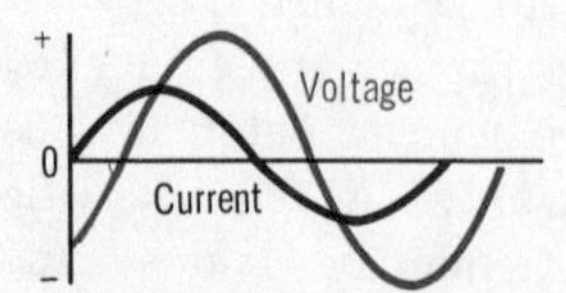

Below resonance, impedance is capacitive, so current leads voltage

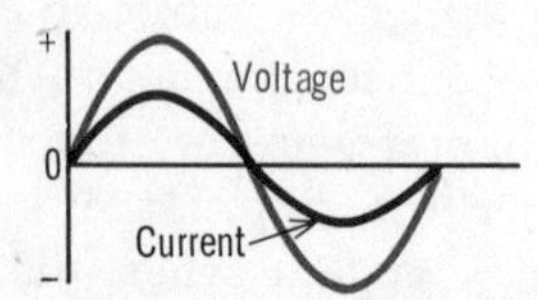

At resonance, impedance equals resistance, so current and voltage are in phase

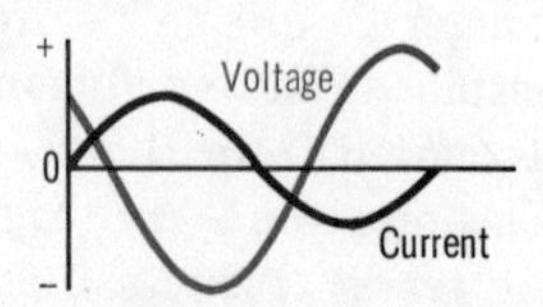

Above resonance, impedance is inductive, so voltage leads current

the resonance band

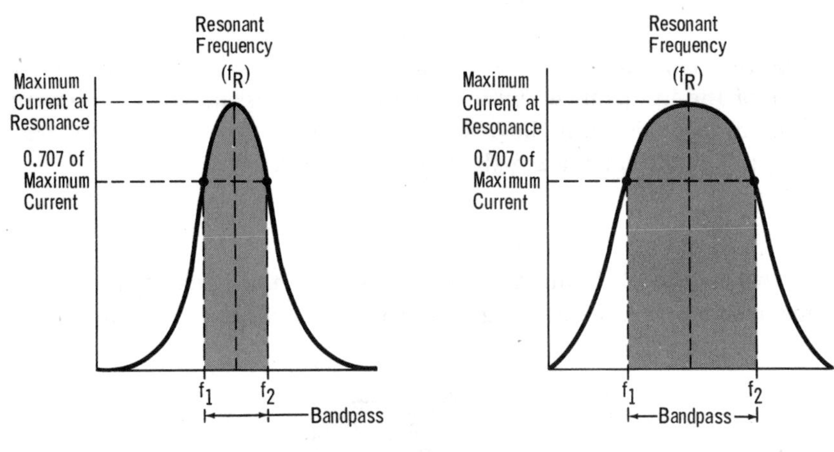

The range of frequencies at which the current has a value of 0.707 or greater times its value at the resonant frequency is called the resonance band, or bandpass

From the resonance curve shown on the previous page, you can see that although maximum current occurs only at the resonant frequency, there is a small *range* of frequencies on either side of resonance where the current is almost as large as it is at resonance. For practical purposes, within this range, or band, of frequencies, the circuit is at resonance. This band of frequencies is called the *resonance band,* or *bandpass,* of the circuit, and serves as a basis for rating or comparing circuits that are used for their resonant properties. Normally, the bandpass is considered as that range of frequencies between which the circuit current is 0.707, or greater, times its value at resonance. Thus, if the current in a particular circuit is 2 amperes at resonance, the bandpass of the circuit is that portion of the resonance curve between the two points corresponding to 1.414 amperes (0.707×2). You can see from this that the width of the bandpass depends on the *shape* of the resonance curve. Curves, which because of their shape have a *wide separation* between the points corresponding to 0.707 of resonance current, have a *wide bandpass.* Those with a *small separation* between the 0.707 points have a *narrow bandpass.*

The resonance band or bandpass of a circuit is also often referred to as the *bandwidth.* All three terms mean the same thing.

effect of resistance on resonance band

The resonant frequency of a series LCR circuit depends only on the values of the inductance and capacitance. The resistance of the circuit has nothing to do with the resonant frequency. This is obvious from the equation for calculating the resonant frequency:

$$f_R = \frac{1}{2\pi\sqrt{LC}}$$

Any two values of L and C whose product is the same will have the same resonant frequency, regardless of the resistance of the circuit.

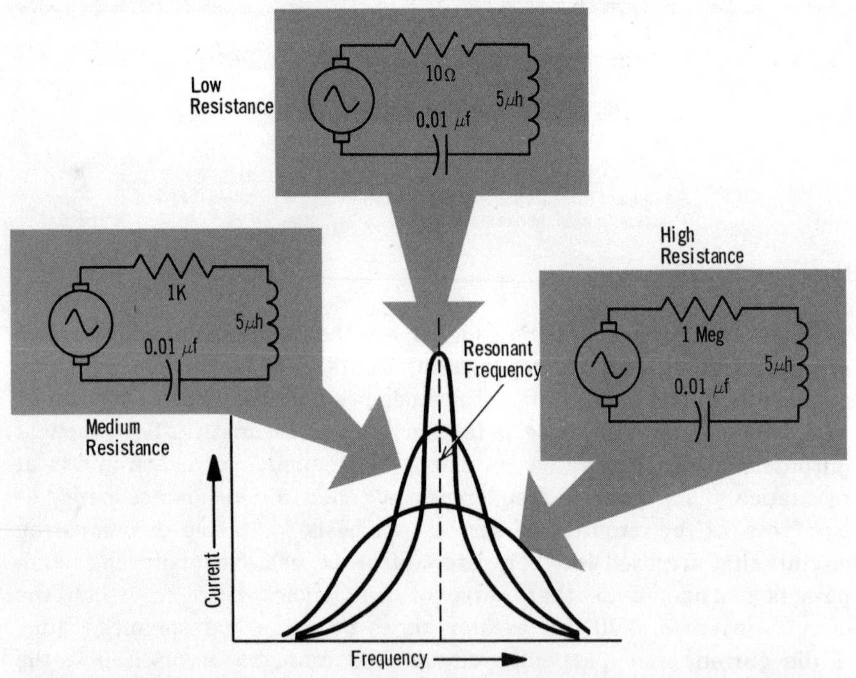

The circuit resistance determines the height and flatness of the resonance curve

Although the resistance plays no part in determining the resonant frequency, it does affect the current that flows at resonance. Since the impedance at resonance is equal to the circuit resistance, resonant current is limited only by the values of the resistance. If the resistance is small, very heavy current flows at resonance. And if the resistance is large, the current will be small, even though the resonant frequency is the same.

effect of resistance on resonance band (cont.)

Since the resonance curve of a series LCR circuit shows how the current varies with frequency, the shape of the curve depends on the value of the circuit resistance. As shown, the greater the resistance, the *lower* is the maximum height of the curve. The reason for this is that the high point on the curve corresponds to the current at resonance. In addition, the greater the resistance, the *flatter* is the curve. The reason for this is that off resonance, the shape of the curve, or the circuit current, is controlled by both the net reactance and the resistance. The net reactance depends on the frequency, but the resistance remains the same for all frequencies. So the higher the resistance, the greater is the relative control it has on the current, and the more it tends to limit the current to a constant value.

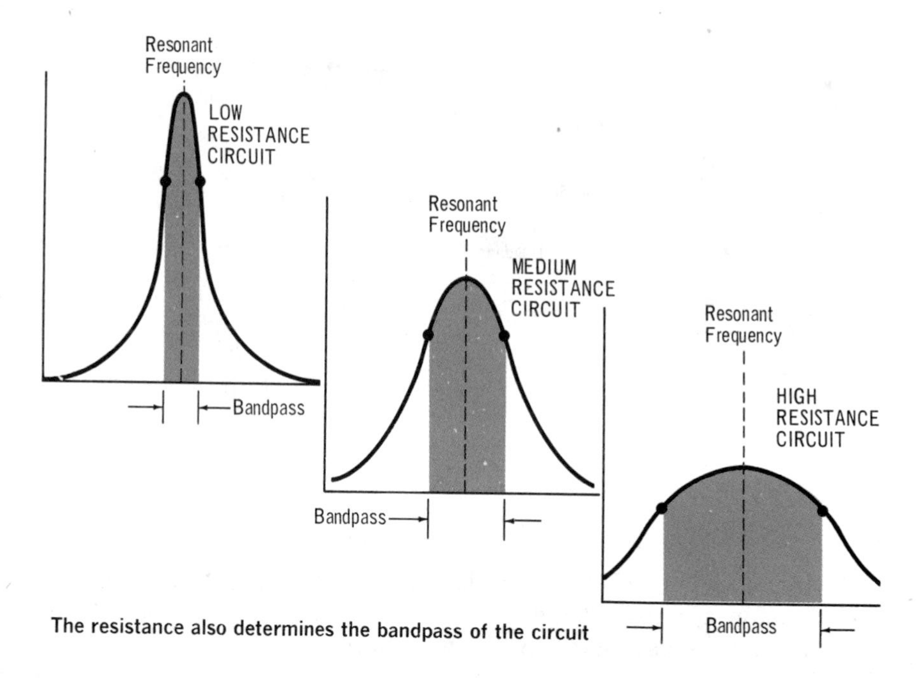

The resistance also determines the bandpass of the circuit

The flatter a resonance curve is, the further apart are the points corresponding to 0.707 of peak, or resonant, current. This means that the flatter the curve, the wider the bandpass; and since resistance determines the flatness of the resonance curve, the *higher* the *resistance,* the *wider* the *bandpass.* Similarly, the *smaller* the *resistance,* the *narrower* the *bandpass.*

quality

At resonance, the voltage drop across the resistance of a series resonant circuit equals the applied voltage, while the voltage drops across the inductance and capacitance are usually many times higher. The ratio of the voltage drop at resonance across either the inductance or capacitance to that across the resistance is used to express the *quality* of a series resonant circuit. This ratio is called the Q of the circuit, and is equal to either E_L/E_R or E_C/E_R. Since E_L and E_C are equal at resonance, either voltage can be used to find the Q of a circuit. If E_L is used, the equation $Q = E_L/E_R$ can also be written in the form $Q = IX_L/IR$, which can then be reduced to what is the standard equation for Q.

$$Q = X_L/R$$

You can see from this equation, that the lower the resistance, the higher the Q, or quality, of the resonant circuit.

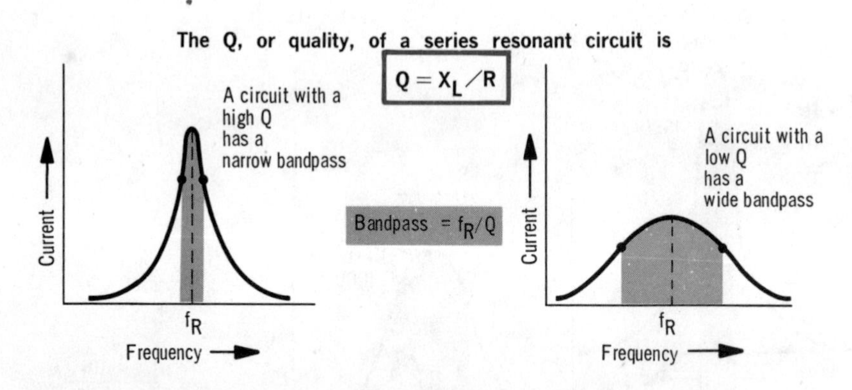

Notice that the equation for the Q of the series resonant LCR circuit is the same as you learned earlier in this volume for the Q of a coil. This shows why the d-c resistance of a coil is important in resonant circuits

A circuit with a high Q has a high, steeply sloping resonance curve, and therefore a narrow bandpass. If the Q of a circuit is known, the bandpass can be calculated by the equation:

$$\text{Bandpass (Hz)} = f_R/Q$$

The result of the equation is the total number of cycles above and below the resonant frequency that make up the resonance band. For example, if for a resonant frequency of 100 Hz this equation gives a result of 10 Hz, the resonance band extends from 95 Hz (100 – 5) to 105 Hz (100 + 5).

practical uses

You may wonder at this point whether series resonant circuits serve any practical use. The answer is yes; not only are they used in practical applications, but they are used extensively. Practically all applications of series resonant circuits make use of their property of allowing a large current to flow at frequencies in the resonance band, and providing a high opposition to current flow at frequencies outside of the resonance band. They are thus used because of their ability to *discriminate* against frequencies outside of the resonance band. As shown, a series combination of inductance, capacitance, and resistance can be connected in a circuit in such a way that it causes the circuit to respond only to those frequencies within the resonance band. Or, the same LCR combination can be connected so that the circuit responds only to those frequencies outside of the resonance band. The operation of the LCR combination is *identical* in both cases, but its effect on the overall circuit is different.

Another use of series resonant circuits is to achieve a *gain in voltage*. You know that at resonance, the voltage across X_L and X_C can be many times greater than the applied voltage. In certain applications, the voltage across either X_L or X_C is used as an output voltage to perform some function. Since this voltage is larger than the applied voltage, which is the input voltage, a voltage gain has been accomplished in the circuit.

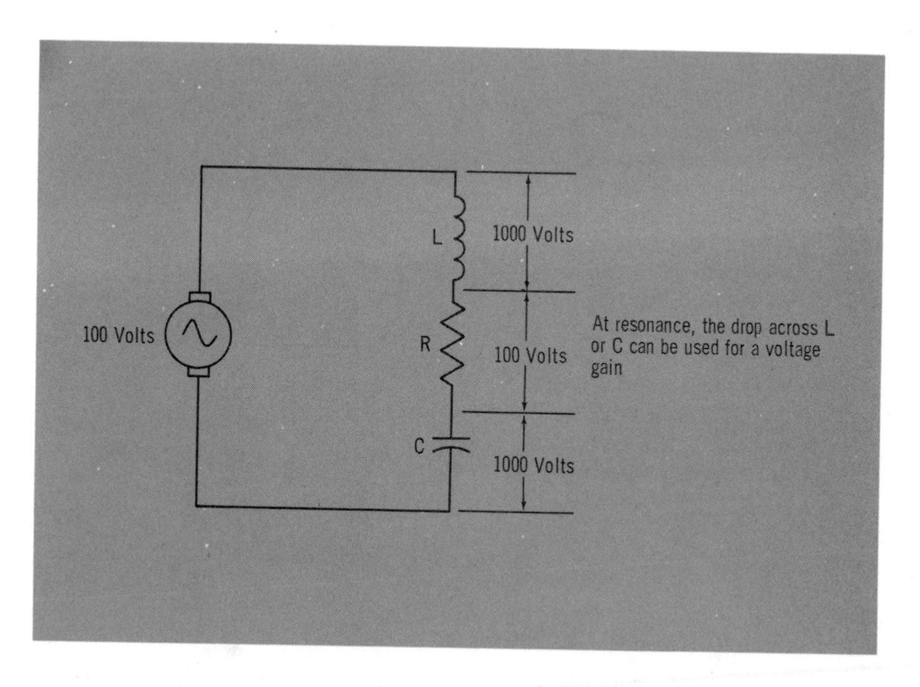

practical uses (cont.)

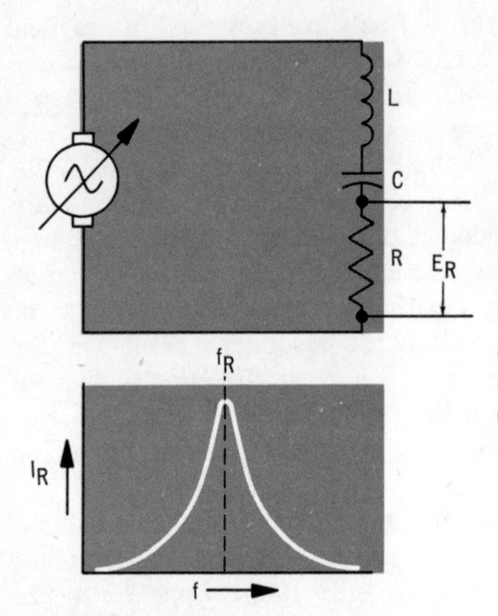

In this circuit, the current through the resistance, I_R, produces the output voltage. At resonance, X_L cancels X_C, so the circuit current, which is also I_R, is maximum. Above and below resonance, the circuit impedance increases, and I_R drops sharply. Thus, only the frequencies at or around resonance produce a useful output voltage

In this circuit, the voltage across R_3, E_{R3}, produces the output. At resonance, the reactances of L and C are opposite and equal, and effectively cancel each other. So the impedance of the LCR circuit is 10 ohms. This makes the parallel combination of R_2 and R_3 10 ohms. Most of the voltage is then dropped across R_1. Off resonance, though, the impedance of the LCR circuit can be much higher than 1000 ohms, so R_3 can drop much more than R_1. As a result, only the frequencies away from resonance will produce a useful output voltage

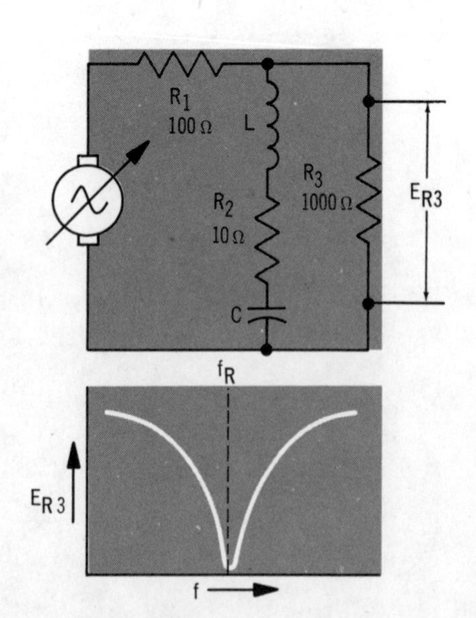

solved problems

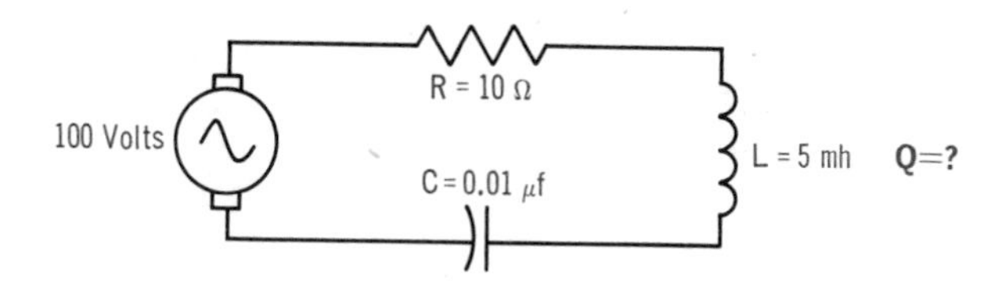

Problem 26. What is the Q of the circuit?

The Q can be calculated from the inductive reactance and resistance, using the equation $Q = X_L/R$. Or it can be found from the voltage drop across L or C and the drop across R, using either the equation $Q = E_L/E_R$ or $Q = E_C/E_R$. No matter which method is used, though, one of the reactances has to be found first; and to do this, you have to know the resonant frequency (f_R) of the circuit. Depending on the method you choose for finding Q, the following are the calculations you have to perform:

If you use $Q = X_L/R$
(1) Find f_R
(2) Find X_L
(3) Find Q

If you use $Q = E_L/E_R$
(1) Find I
(2) Find f_R
(3) Find X_L
(4) Find E_L
(5) Find Q

You can see from this comparison that the problem can be solved more easily using the equation $Q = X_L/R$.

Calculating the Resonant Frequency, f_R*:*

$$f_R = \frac{1}{2\pi\sqrt{LC}} = \frac{1}{6.28 \times \sqrt{0.005 \times 0.00000001}} = 22{,}500 \text{ Hz} = 22.5 \text{ kHz}$$

Calculating X_L: The inductive reactance at the resonant frequency (f_R) is the value to be calculated.

$$X_L = 2\pi f_R L = 6.28 \times 22{,}500 \times 0.005 = 707 \text{ ohms}$$

Calculating the Q:

$$Q = X_L/R = 707/10 = 70.7$$

Problem 27. What is the bandwidth of the circuit?

Since the Q of the circuit is known, the bandwidth can be easily calculated from the equation:

$$\text{Bandpass (Hz)} = f_R/Q = 22{,}500/70.7 = 318 \text{ Hz}$$

This means that for practical purposes the circuit can be considered as resonant at frequencies between 22,341 and 22,659 Hz.

solved problems (cont.)

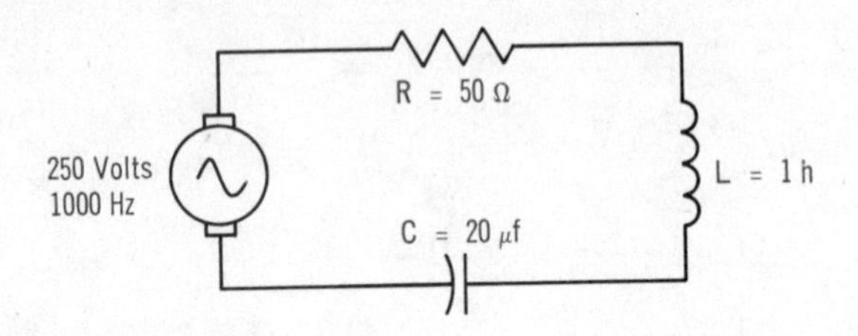

Problem 28. Is the circuit inductive or capacitive?

A circuit is inductive if X_L is larger than X_C, and capacitive if X_C is larger than X_L. So you could solve this problem by calculating X_L and X_C, and seeing which is larger. This would require two separate calculations. The problem can be solved with one calculation if you remember that when the frequency of a series LCR circuit is below the resonant frequency, the circuit is capacitive; and if the frequency is above the resonant frequency, the circuit is inductive. So all you have to do is calculate the resonant frequency and compare it to the actual circuit frequency. The resonant frequency, then, is

$$f_R = \frac{1}{2\pi\sqrt{LC}} = \frac{1}{6.28 \times \sqrt{1 \times 0.00002}} = \frac{1}{0.028} = 38 \text{ Hz}$$

The circuit frequency of 1000 Hz is higher than the resonant frequency of 38 Hz, so the circuit is *inductive.*

Problem 29. What is the current in the circuit?

The current is found using the same methods that apply to any series LCR circuit. This means that X_L and X_C must first be found, then the impedance, Z, and finally the current with the equation $I = E/Z$.

Calculating X_L and X_C:

$$X_L = 2\pi fL = 6.28 \times 1000 \times 1 = 6280 \text{ ohms}$$

$$X_C = \frac{1}{2\pi fC} = \frac{1}{6.28 \times 1000 \times 0.00002} = 8 \text{ ohms}$$

Calculating Z:

$$Z = \sqrt{R^2 + (X_L - X_C)^2} = \sqrt{(50)^2 + (6280 - 8)^2} = 6272 \text{ ohms}$$

Notice here that the impedance has almost exactly the same value as X_L. The reason for this is that X_L is so much larger than X_C that the effect of X_C is negligible. In practical work, if X_L is more than 10 times larger than X_C, the value of X_C could actually be neglected, and the impedance considered equal to X_L.

Calculating I:

$$I = E/Z = 250 \text{ volts}/6272 \text{ ohms} = 0.040 \text{ ampere}$$

summary

☐ Series resonance in a series LCR circuit occurs at the frequency at which the net reactance is zero. ☐ At resonance, the impedance is minimum, and current is maximum. ☐ The resonant frequency is $f_R = 1/(2\pi\sqrt{LC})$. ☐ At resonance, the series LCR circuit is completely resistive. ☐ Above resonance, X_L is greater than X_C, and the circuit is inductive. ☐ Below resonance, X_C is greater than X_L, and the circuit is capacitive. ☐ The resonance curve indicates values of current at different frequencies.

☐ The bandpass of a series LCR circuit is that range of frequencies between which the circuit current is 0.707, or greater, times its values at resonance. ☐ The resonance band, or bandpass, of a circuit is also called the bandwidth. ☐ Resistance in a series LCR circuit affects the flatness of the bandpass. The higher the resistance, the wider the bandpass; the smaller the resistance, the narrower the bandpass.

☐ The Q of a circuit is the ratio of the voltage drop across the inductor or capacitor at resonance to the voltage across the resistor. $Q = E_L/E_R = E_C/E_R$. Other equations for Q are $Q = X_L/R$ and $Q = X_C/R$. ☐ The bandpass of a series LCR circuit is given by: Bandpass $= f_R/Q$. ☐ The deviation above and below the resonant frequency is 1/2 the bandpass.

review questions

1. What is the resonant frequency of a series LCR circuit having an L of 15 μh, a C of 15 μf, and an R of 10 ohms?
2. For the circuit of Question 1, what is the value of Q?
3. Derive the equation for the bandpass of a series LCR circuit in terms of the resistance and inductance. (*Hint:* Substitute for Q in the bandpass equation.)
4. What is the bandwidth of a series LCR circuit with R equal to 6.28 ohms and L equal to 50 millihenrys?
5. What is the power factor for a series resonant circuit?
6. If 50 volts is applied to a series LCR circuit with R equal to 5 ohms, X_L equal to 23.2 ohms, and X_C equal to 49.6 ohms, what current will flow at resonance?
7. If 50 volts is applied to a series LCR circuit whose Q is 10, what is the voltage across the capacitor at resonance?
8. In a series LCR circuit, what determines the flatness of the bandpass curve?
9. At resonance, is a series LCR circuit inductive?
10. A series LCR circuit has a current of 20 amperes at its resonant frequency of 100 kHz. What is the bandwidth if at 104 kHz the current is 14.14 amperes?

parallel LC circuits

A parallel LC circuit consists of an inductance and a capacitance connected in parallel across a voltage source. The circuit thus has two branches: an *inductive branch,* and a *capacitive branch.* In an *ideal* parallel LC circuit, which we will consider here, there is *no resistance* in either branch. This is of course impossible; but in actual practice, the resistance can be made so small as to be negligible.

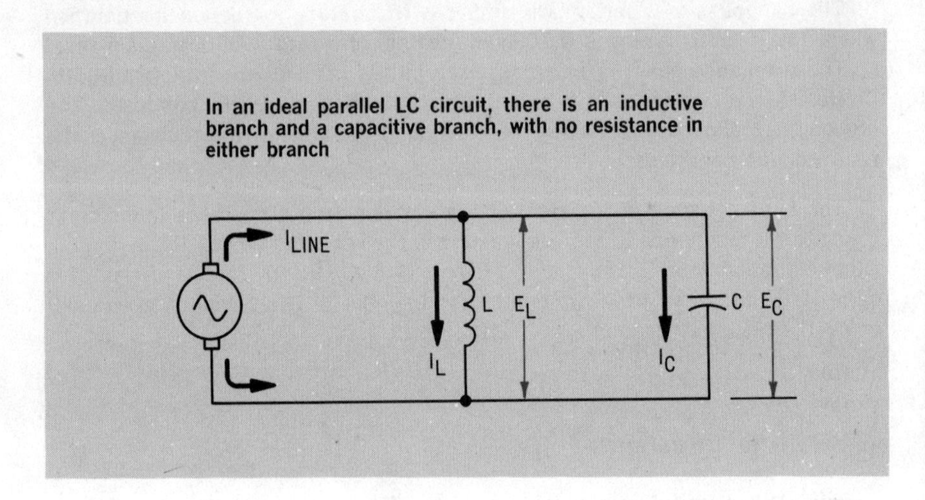

Parallel LC circuits can have more than one inductive or capacitive branch, as well as more than one of both. However, once these circuits are reduced to their equivalent two-branch circuit, their analysis is the same as that of a simple parallel LC circuit.

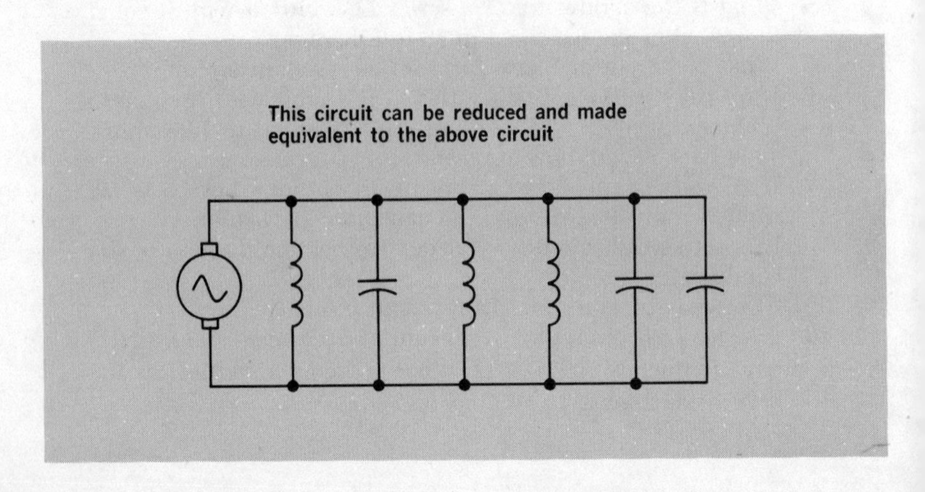

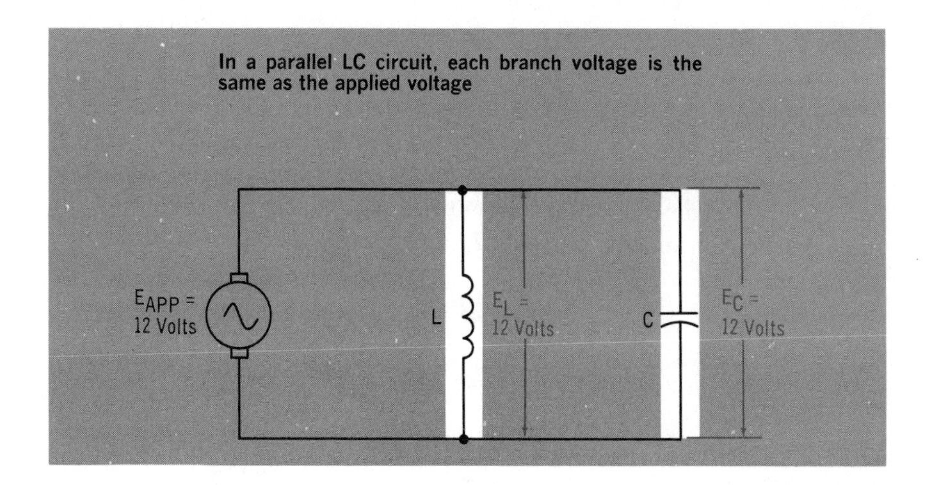

voltage

The voltages across the branches of a parallel LC circuit are the same as the applied voltage, as they are in all parallel circuits. Since they are actually the *same voltage,* the branch voltages and source voltage are equal to each other and in phase. Because of this, the voltage is used as the zero-degree *phase reference,* and the phases of the other circuit quantities are expressed in relation to the voltage. The amplitude of the voltage in a parallel LC circuit is related to the circuit impedance and the line current by Ohm's Law. Thus,

$$E = I_{LINE}Z$$

Since all the circuit voltages are the same, the voltage is used as the zero-degree reference

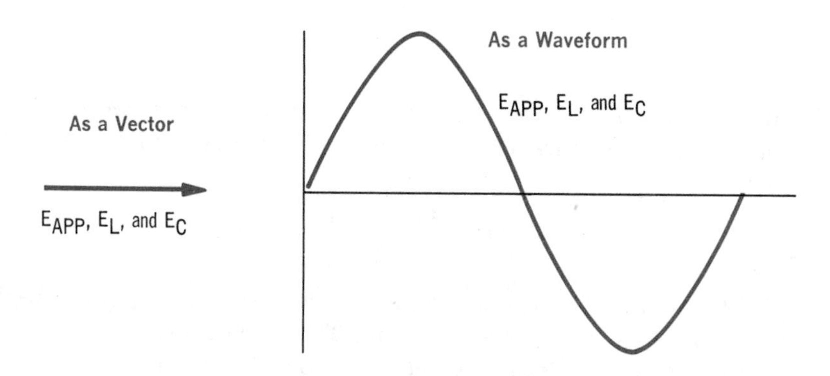

current

The currents in the branches of a parallel LC circuit are *both* out of phase with the circuit voltage. The current in the inductive branch (I_L) lags the voltage by 90 degrees, while the current in the capacitive branch (I_C) leads the voltage by 90 degrees. Since the voltage is the same for both branches, currents I_L and I_C are, therefore, 180 degrees

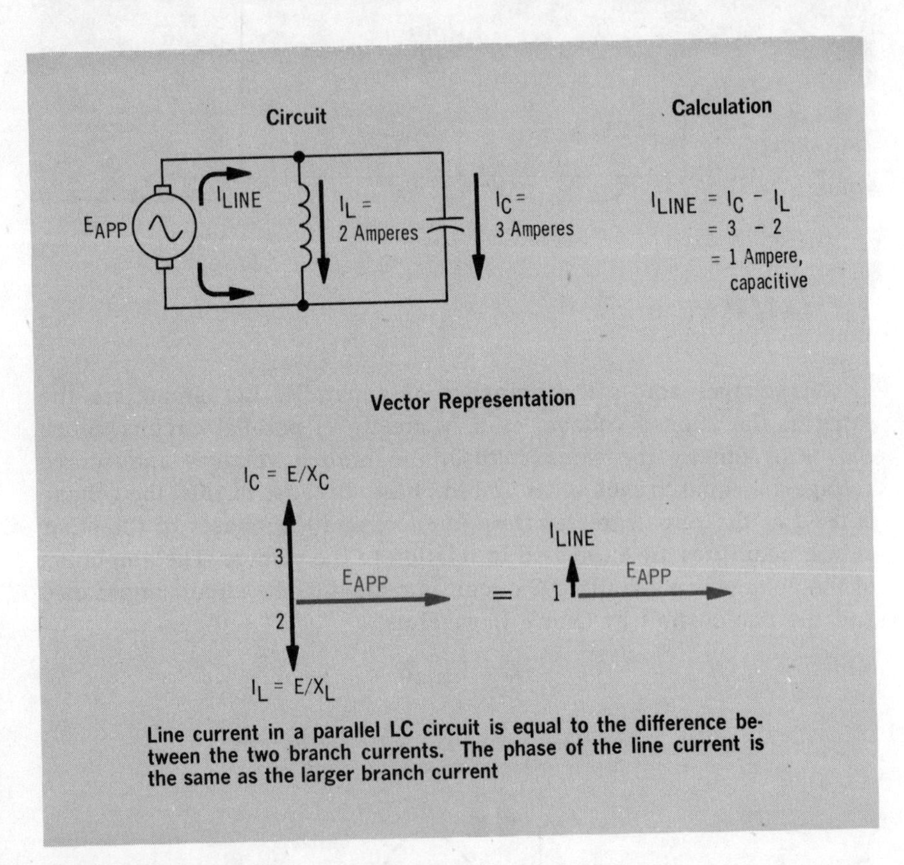

Line current in a parallel LC circuit is equal to the difference between the two branch currents. The phase of the line current is the same as the larger branch current

out of phase. The amplitudes of the branch currents depend on the value of the reactance in the respective branches, and can be found from:

$$I_L = E/X_L$$

and

$$I_C = E/X_C$$

With the branch currents being 180 degrees out of phase, the line current is equal to their vector sum. This vector addition is done by subtracting the smaller branch current from the larger.

current (cont.)

The line current for a parallel LC circuit, therefore, has the *phase characteristics* of the *larger* branch current. Thus, if the inductive branch current is the larger, the line current is inductive, and lags the applied voltage by 90 degrees. And if the capactive branch current is the larger, the line current is capacitive, and leads the applied voltage by 90 degrees. In equation form, therefore, the line current is

$$I_{LINE} = I_L - I_C \text{ (if } I_L \text{ is larger than } I_C)$$

$$I_{LINE} = I_C - I_L \text{ (if } I_C \text{ is larger than } I_L)$$

If the impedance of the circuit is known, the line current can also be found by Ohm's Law:

$$I_{LINE} = E/Z$$

A unique property of the line current in a parallel LC circuit is that it is always *less* than one of the branch currents, and sometimes less than both. This is in contrast to all other parallel circuits you have studied, in which the line current was always greater than any one of the branch currents. The reason that the line current is less than the branch currents is because the two branch currents are 180 degrees out of phase. As a result of the phase difference, some *cancellation* takes place between the two currents when they combine to produce the line current. You will find later that this property is the basis of *parallel resonance*.

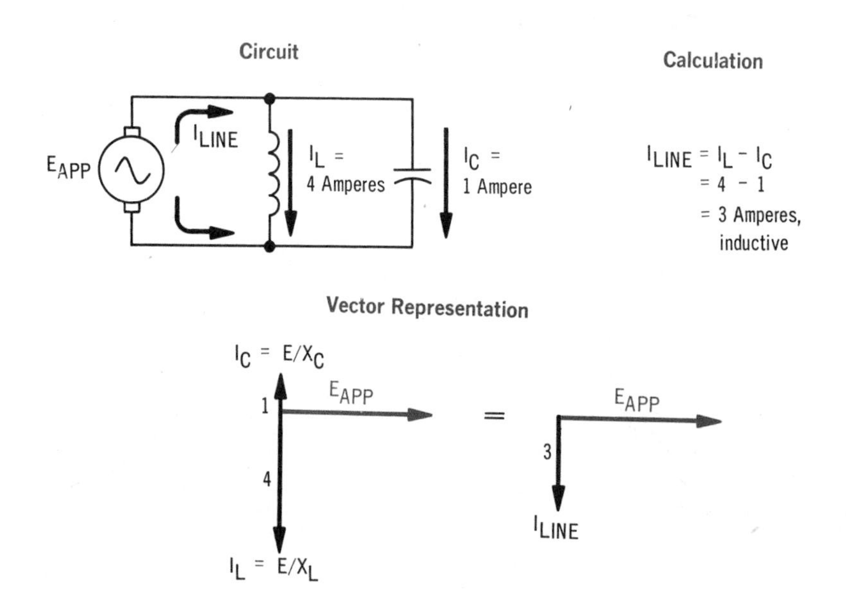

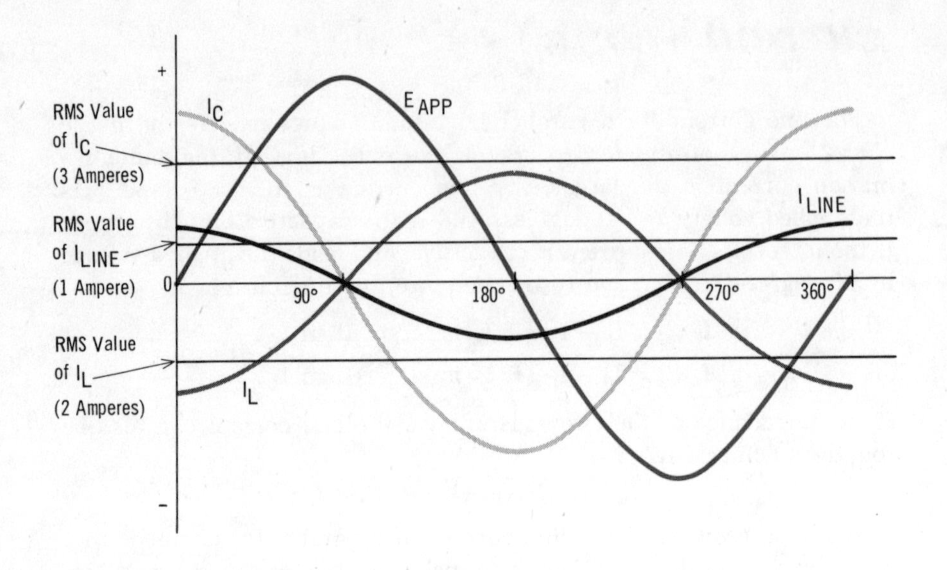

current waveforms

The waveforms of the currents in a parallel LC circuit are similar to the voltage waveforms you have seen for a series LC circuit. All of the instantaneous values of two waveforms 180 degrees out of phase are added to produce the resulting waveform, which in this case is the line current waveform. The current waveforms for the two circuits solved on the previous pages are shown.

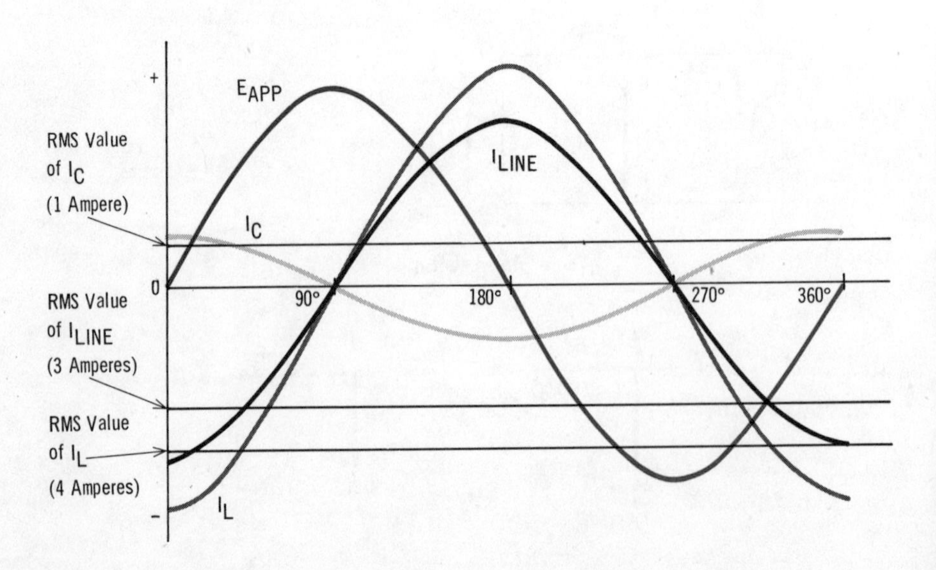

impedance

The impedance of a parallel LC circuit can be found with:

$$Z = \frac{X_L \times X_C}{X_L - X_C} \quad \text{(for } X_L \text{ larger than } X_C\text{)}$$

and

$$Z = \frac{X_L \times X_C}{X_C - X_L} \quad \text{(for } X_C \text{ larger than } X_L\text{)}$$

When using these equations, the impedance will have the phase characteristic of the *smaller* reactance.

For mathematical simplicity, a single equation can be used to find Z regardless of whether X_L or X_C is larger:

$$Z = \frac{X_L \times X_C}{X_L + X_C}$$

This is the same equation used for parallel resistances, but since X_L and X_C are 180 degrees out of phase, to use this equation, X_L is always a positive (+) quantity, and X_C is always a negative (−) quantity. When the *relative values* of X_L and X_C make Z *negative*, the impedance is *capacitive*. Similarly, when Z is *positive*, the impedance is *inductive*. Remember that X_C is not *actually* a negative quantity. It is assumed so only for this impedance equation.

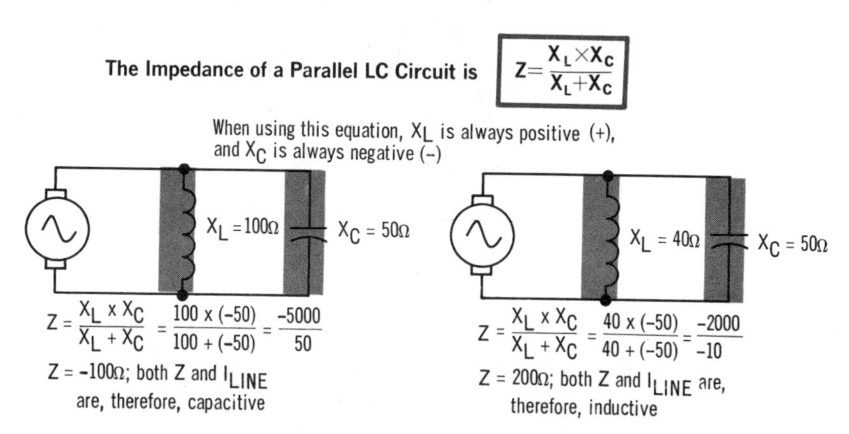

You can see that because of the difference in sign, the *closer* X_L and X_C are in value, the *larger* is the impedance. When X_L and X_C are equal, the impedance is infinitely large. As you will learn later, the circuit is then at resonance.

If the line current and applied voltage are known, the impedance can also be found by Ohm's Law:

$$Z = E_{APP}/I_{LINE}$$

effect of frequency

The frequency of the applied voltage affects the values of X_L and X_C in a parallel LC circuit. Since the value of the impedance is based on X_L and X_C, the frequency also affects the impedance. But because X_L and X_C change in *opposite* directions for a given change in frequency, no general relationships can be given for the effects of frequency on impedance, as was done for RL and RC circuits. However, as was mentioned previously in series LC circuits, for every combination of inductance and capacitance, there is one frequency called the *resonant frequency,* where the value of X_L *equals* that of X_C. And, as you learned on the previous page, when X_L and X_C are equal in a parallel LC circuit, the impedance approaches an infinitely high value. At frequencies above and below this resonant frequency, X_L and X_C have different values, and the impedance is lower.

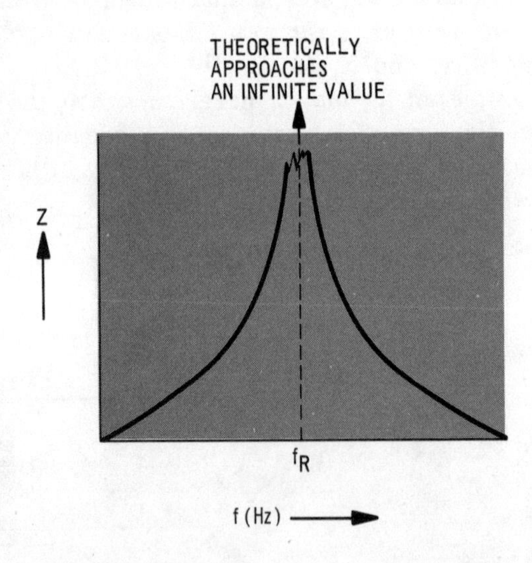

The frequency response curve of a parallel LC circuit would look like this. The high, or infinite impedance, point on the curve occurs at the resonant frequency

If the frequency changes away from this point, the impedance decreases. And if the frequency changes toward this point, the impedance increases

It can be said, therefore, that in parallel LC circuits, a change in frequency *towards* the resonant frequency will cause the *impedance* to *increase* and the *line current* to *decrease*. Similarly, a change in frequency away from the resonant frequency causes the *impedance* to *decrease* and the *line current* to *increase*.

solved problems

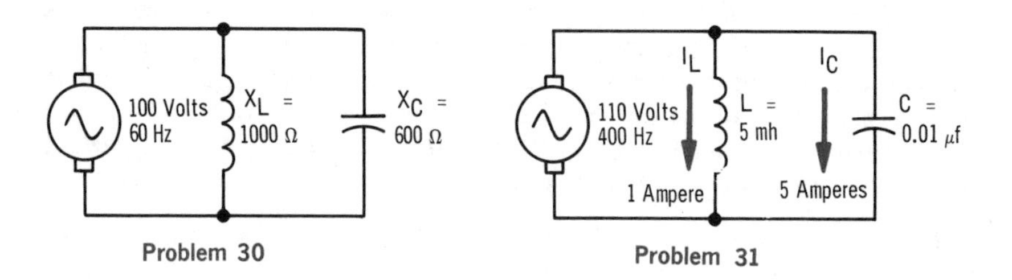

Problem 30. What is the line current in the circuit?

The problem can be solved in either of two ways. One is by finding the two branch currents, and then using them to calculate the line current. The other is by finding the impedance, and then using Ohm's Law to determine the line current. Both methods are equally suitable, so we will solve the problem both ways.

Finding I_{LINE} By the Branch Currents:

$$I_L = E_{APP}/X_L = 100/1000 = 0.1 \text{ ampere}$$
$$I_C = E_{APP}/X_C = 100/600 = 0.166 \text{ ampere}$$

Since I_C is larger than I_L, the equation for I_{LINE} is

$$I_{LINE} = I_C - I_L = 0.166 - 0.1 = 0.066 \text{ ampere}$$

I_C was the larger branch current, so the line current of 0.066 ampere is capacitive, which means it leads the applied voltage by 90 degrees.

Finding I_{LINE} By the Impedance:

$$Z = \frac{X_L \times X_C}{X_L + X_C} = \frac{1000 \times (-600)}{1000 + (-600)} = -1500 \text{ ohms, capacitive}$$

The line current, therefore, is

$$I_{LINE} = 100/1500 = 0.066 \text{ ampere}$$

Since the impedance was negative, the line current is capacitive, and so leads the applied voltage by 90 degrees.

Problem 31. In the circuit, which is larger? X_L or X_C?

This problem could be solved by calculating the values of X_L and X_C, and comparing them. But since you only have to determine which is larger, the problem can be solved by inspection. The voltages across both branches are the same, so the larger reactance will allow the least branch current to flow. Since I_L (1 ampere) is smaller than I_C (5 amperes), X_L must be larger than X_C.

summary

☐ Because the voltage across each branch of a parallel LC circuit is the same, the voltage is used as the phase reference. ☐ The inductive branch current is found by: $I_L = E/X_L$; and the capacitive branch current by: $I_C = E/X_C$. ☐ The line current has the characteristic of the larger branch current, and is equal in magnitude to the difference of the two currents: $I_{LINE} = I_L - I_C$, for I_L larger than I_C; and $I_{LINE} = I_C - I_L$, for I_C larger than I_L. ☐ The line current can also be found by Ohm's Law: $I_{LINE} = E/Z$.

☐ The impedance of a parallel LC circuit is $Z = X_L X_C/(X_L - X_C)$, for X_L greater than X_C; or $Z = X_L X_C/(X_C - X_L)$, for X_C greater than X_L. ☐ The parallel LC circuit is either capacitive or inductive, depending on the smaller of the two parallel reactances. ☐ Impedance in a parallel LC circuit can also be found by: $Z = X_L X_C/(X_L + X_C)$. When using this equation, X_C is always assigned a negative (−) sign. The resulting sign of Z then determines whether the impedance is inductive (+) or capacitive (−).

☐ At the resonant frequency, where X_L equals X_C, the impedance approaches an infinitely high value. ☐ At frequencies above and below the resonant frequency, X_L and X_C have different values, and the impedance is lower than at resonance. ☐ At resonance, the line current in a parallel LC circuit will be minimum, and will increase as the frequency is changed above and below resonance.

review questions

1. The line current in a parallel LC circuit is 20 amperes, and the current through the inductor is 30 amperes. What is the current through the capacitor?
2. If the current through the inductor in Question 1 is 15 amperes, what is the current through the capacitor?
3. For a parallel LC circuit, X_L is 20 ohms, and X_C is 10 ohms. What is Z?
4. Answer Question 3, where the values for X_L and X_C are interchanged.
5. Is the circuit of Question 3 inductive or capacitive?
6. Is the circuit of Question 4 inductive or capacitive?
7. Can the line current of a parallel LC circuit ever be greater than either or both branch currents?
8. At what frequency will an ideal parallel LC circuit approach an open circuit?
9. What is the line current for the circuit of Question 8?
10. Will an increase in frequency above the resonant frequency in a parallel LC circuit cause the circuit to become more capacitive or more inductive?

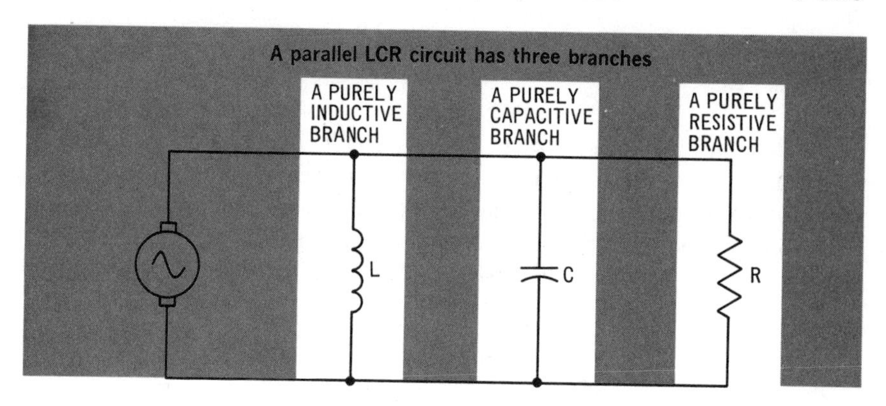

parallel LCR circuits

A parallel LCR circuit is essentially a parallel LC circuit having a resistance in parallel with the inductance and capacitance. There are thus *three branches* in the circuit: a purely inductive branch, a purely capacitive branch, and a purely resistive branch. You have already learned how to analyze and solve parallel circuits that contain any two of these branches. You will now learn how to analyze circuits that have all three.

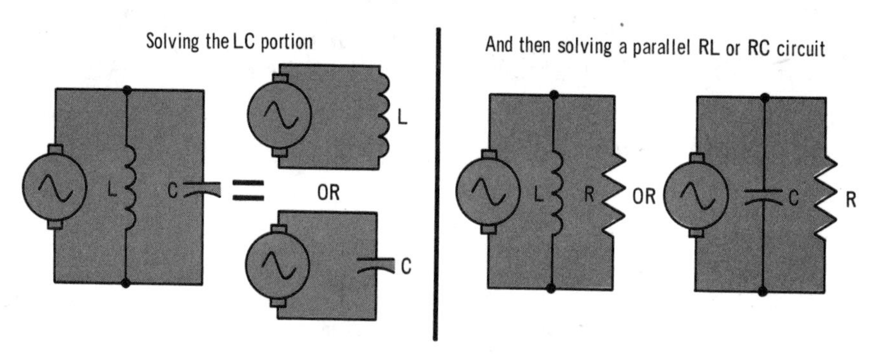

Actually, you will find that the solution of a parallel circuit is basically the solution of a parallel LC circuit, and then the solution of either a parallel RL circuit or a parallel RC circuit. The reason for this is that, as you will recall from the previous pages, a parallel combination of L and C appears to the source as a pure L or a pure C. So by solving the LC portion of a parallel LCR circuit first, you, in effect, reduce the circuit to an equivalent RL or RC circuit.

voltage

The distribution of the voltage in a parallel LCR circuit is no different than it is in a parallel LC circuit, or in any parallel circuit. The branch voltages are all *equal* and *in phase*, since they are the same as the applied voltage. The resistance is simply another branch across which the applied voltage appears. Because the voltages throughout the circuit

In a parallel LCR circuit, each branch voltage is the same as the applied voltage

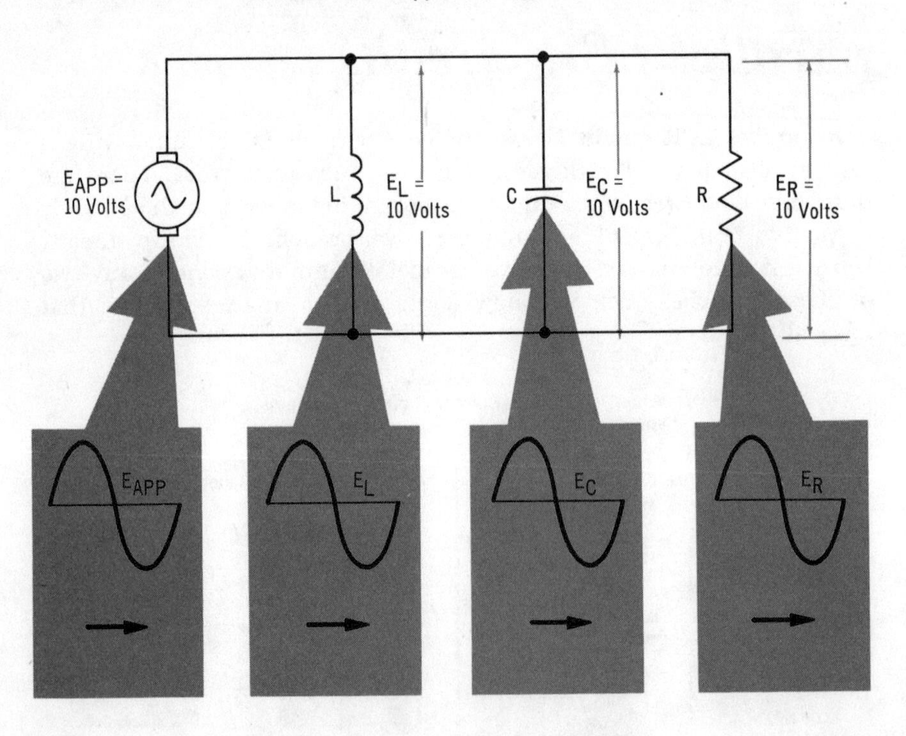

Since the circuit voltages are all the same, the voltage is used as the zero-degree phase reference

are the same, the applied voltage is again used as the zero-degree *phase reference*, as it was in parallel LC circuits. The phase angles of all other circuit quantities, then, are expressed in relation to the applied voltage. The amplitude of the applied voltage is related to the circuit impedance by Ohm's Law:

$$E_{APP} = I_{LINE}Z$$

current

The three branch currents in a parallel LCR circuit are an *inductive current* (I_L), a *capacitive current* (I_C), and a *resistive current* (I_R). Each is independent of the other, and depends only on the applied voltage and the branch resistance or reactance. The amplitudes of the branch currents are equal to:

$$I_L = E_{APP}/X_L \qquad I_C = E_{APP}/X_C \qquad I_R = E_{APP}/R$$

The three branch currents all have *different* phases with respect to the branch voltages. I_L lags the voltage by 90 degrees, I_C leads the voltage by 90 degrees, and I_R is in phase with the voltage. Since the voltages are the same, I_L and I_C are 180 degrees out of phase with each other, and both are 90 degrees out of phase with I_R. Because I_R is in phase with the voltage, it has the same zero-reference direction as the voltage. So I_C leads I_R by 90 degrees, and I_L lags I_R by 90 degrees.

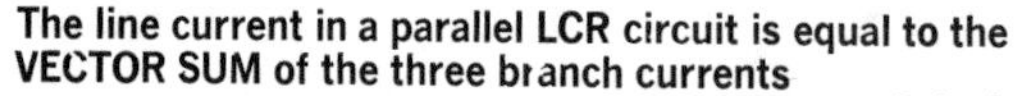
The line current in a parallel LCR circuit is equal to the VECTOR SUM of the three branch currents

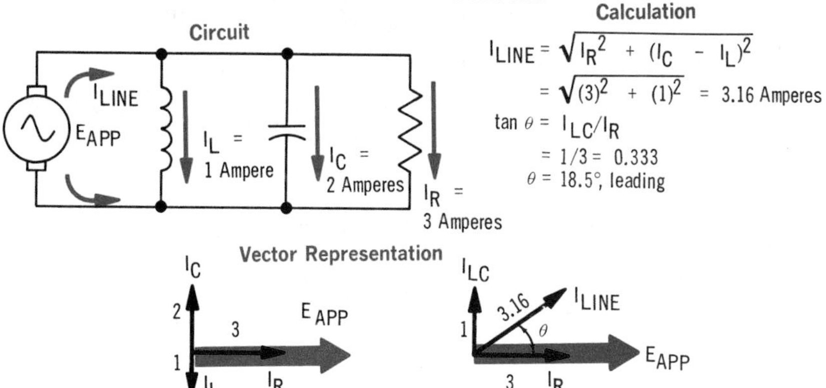

The reactive branch that has the largest current determines whether the line current leads or lags the applied voltage. The relative amplitudes of the total reactive and resistive currents then determine the angle of lead or lag.

The line current (I_{LINE}) is the vector sum of the three branch currents, and so can be calculated by adding I_L, I_C, and I_R vectorially. The different phase relationships between the three branch currents make it necessary to perform this addition in two steps. First, the two reactive currents are added, using the same methods learned for parallel LC circuits. The total of the currents, called I_{LC}, then is

$$I_{LC} = I_L - I_C \qquad \text{(if } I_L \text{ is larger than } I_C\text{)}$$

$$I_{LC} = I_C - I_L \qquad \text{(if } I_C \text{ is larger than } I_L\text{)}$$

current (cont.)

To find the line current, the quantity I_{LC} is then added to I_R, using the Pythagorean Theorem. Therefore,

$$I_{LINE} = \sqrt{I_R^2 + I_{LC}^2}$$

When the equations for these two additions are combined, they give the standard equation for line current in terms of the branch currents. This is

$$I_{LINE} = \sqrt{I_R^2 + (I_L - I_C)^2} \quad \text{(if } I_L \text{ is larger than } I_C\text{)}$$

$$I_{LINE} = \sqrt{I_R^2 + (I_C - I_L)^2} \quad \text{(if } I_C \text{ is larger than } I_L\text{)}$$

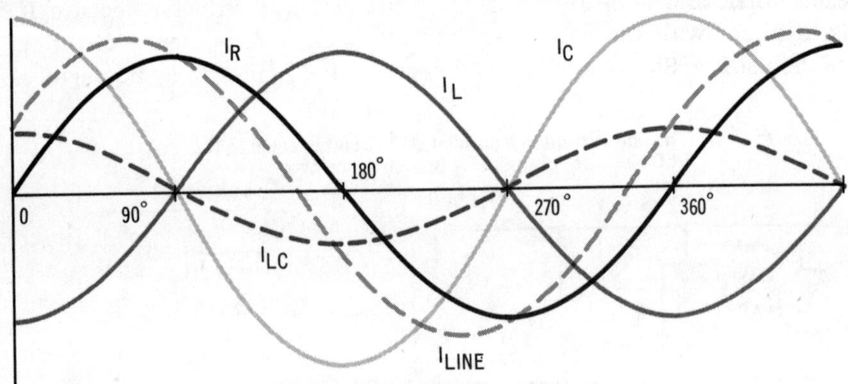

To find the line current (I_{LINE}), first add the inductive current (I_L) to the capacitive current (I_C) algebraically, to find the total reactive current (I_{LC}). Then use the Pythagorean Theorem with I_{LC} and the resistive current, I_R

Whether the line current leads or lags the applied voltage depends on which of the reactive branch currents, I_L or I_C, is the *larger*. If I_L *is larger*, I_{LINE} *lags* the applied voltage; and if I_C *is larger*, I_{LINE} *leads* the applied voltage. The exact angle of lead or lag is found by the equation:

$$\tan \theta = I_{LC}/I_R$$

Whether the angle is leading or lagging depends on which branch current, I_L or I_C, is the larger.

current waveforms

The waveforms of the currents in a parallel LCR circuit are similar to the waveforms for the voltages in a series LCR circuit. The instantaneous values of three out-of-phase waveforms combine to form one resulting waveform, which, in this case, is the circuit line current. Two of the waveforms are 180 degrees out of phase, and so their instantaneous values are always of opposite polarity. The third waveform is 90 degrees out of phase with the other two, but in phase with the applied voltage waveform. Representative waveforms of a parallel LCR circuit are shown.

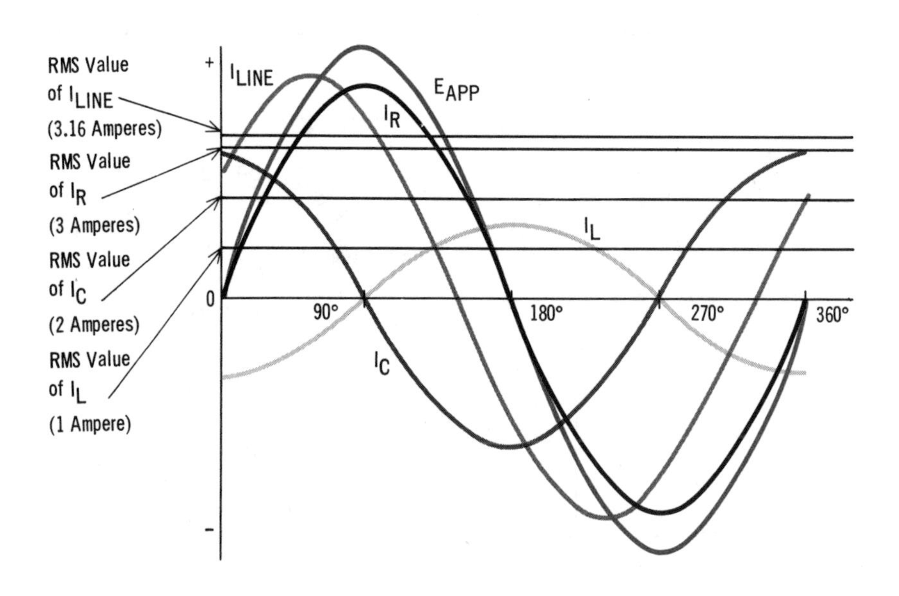

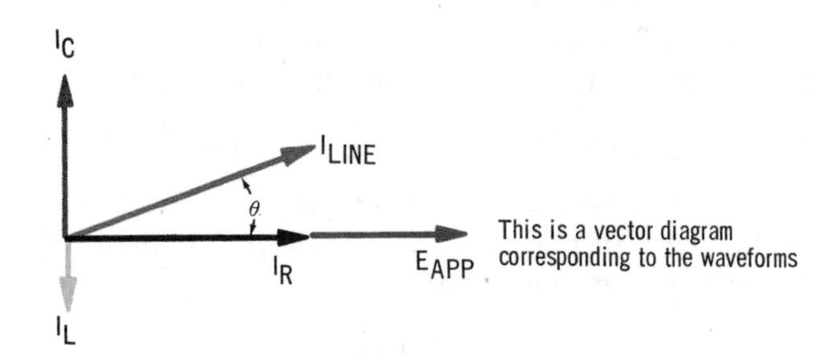

impedance

To determine the impedance of a parallel LCR circuit, you must first find the net reactance (X) of the inductive and capacitive branches. Then, using X, you can find the impedance (Z) the same as you would in a parallel RL or RC circuit:

$$X = \frac{X_L \times X_C}{X_L + X_C}$$

$$Z = \frac{XR}{\sqrt{X^2 + R^2}}$$

Remember that X_L is a positive quantity, and X_C is negative. Therefore, both X and Z will also be either negative (capacitive) or positive (inductive).

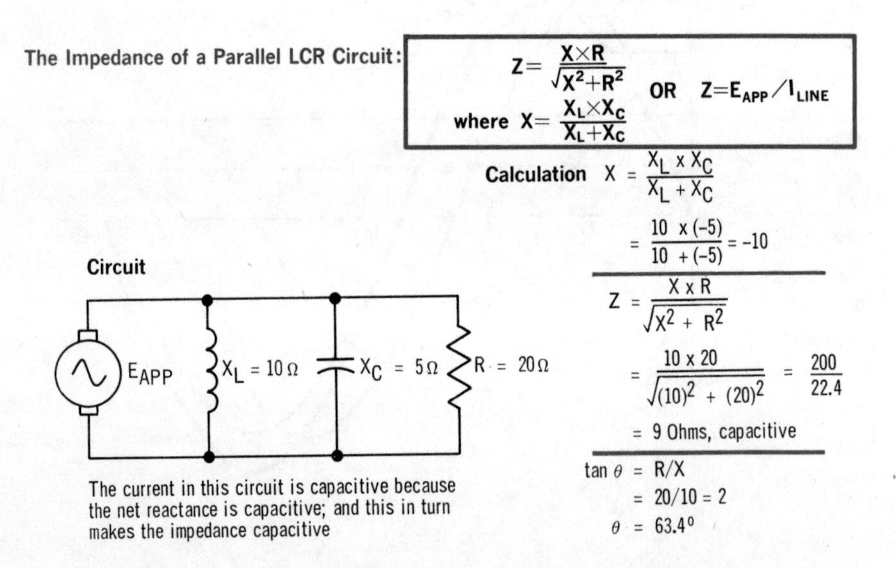

The current in this circuit is capacitive because the net reactance is capacitive; and this in turn makes the impedance capacitive

Whenever Z is inductive, the line current will lag the applied voltage. Similarly, when Z is capacitive, the line current will lead the applied voltage. The exact angle of lead or lag depends on the relative values of X and R. It can be found by the equations:

$$\tan \theta = R/X \qquad \text{or} \qquad \cos \theta = Z/R$$

Again, the 10-to-1 rule given for the other circuits applies here.

If the line current and the applied voltage are known, the impedance can also be found by Ohm's Law:

$$Z = E_{APP}/I_{LINE}$$

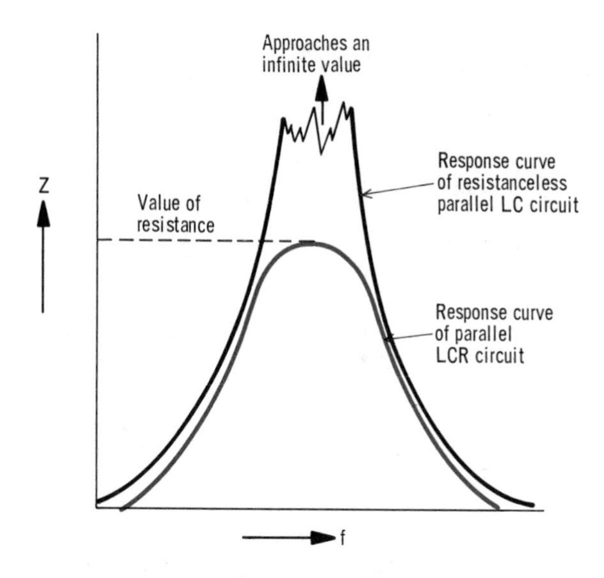

The impedance of a parallel LCR circuit can never be larger than the resistance, no matter how close to the resonant frequency the circuit frequency approaches

effect of frequency

The effects of frequency on parallel LCR circuits are similar to those for parallel LC circuits. A change in frequency causes changes in the values of both X_L and X_C, but in *different* directions. And this, in turn, results in a change in the circuit impedance. However, the exact manner in which the impedance changes depends on the *relative values* of X_L and X_C.

A definite relationship between the frequency and the impedance can only be stated in relation to the *resonant frequency,* which, you recall, is the frequency that results in X_L and X_C having equal values. This relationship is that any change in frequency *towards* the *resonant* frequency causes an *increase* in *impedance;* whereas a change in frequency *away* from the *resonant* frequency results in a *decrease* in *impedance.* This is the same relationship that exists in a parallel LC circuit. However, there is one important difference.

In an LC circuit, there is theoretically no limit to how large the impedance can go as the frequency approaches the resonant frequency. But in an LCR circuit, the impedance can *never* be larger than the value of the resistance. In effect, the resistance destroys an important characteristic of the parallel inductance and capacitance: the ability to present a very high impedance to the voltage source. This will be covered later.

solved problems

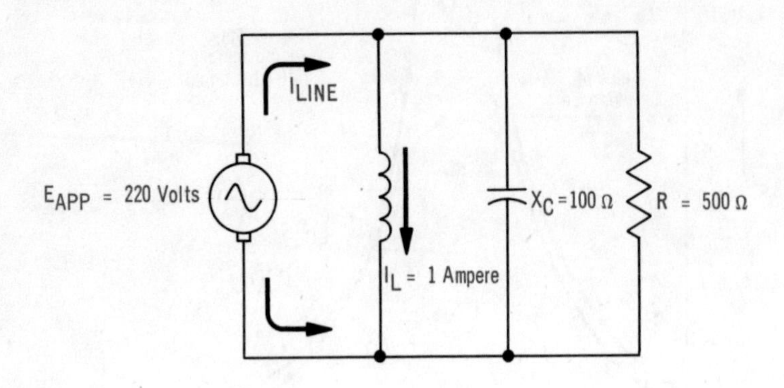

Problem 32. What is the line current in the circuit?

This problem could be solved in two ways. One is by finding all of the branch currents, and using them to calculate the line current. The other is by using the equation $X_L = E_{APP}/I_L$ to find the reactance of the inductive branch, and then calculating the circuit impedance and using it to find the line current. Here, we will use the branch currents to find I_{LINE}.

Calculating I_C and I_R: Since I_L is given, only I_C and I_R have to be found.

$$I_C = E_{APP}/X_C = 220/100 = 2.2 \text{ amperes}$$

$$I_R = E_{APP}/R = 220/500 = 0.44 \text{ ampere}$$

Calculating I_{LINE}:

$$I_{LINE} = \sqrt{I_R^2 + (I_C - I_L)^2} = \sqrt{(0.44)^2 + (1.2)^2} = 1.3 \text{ amperes}$$

The current leads the applied voltage, since I_C is larger than I_L.

Problem 33. What is the phase angle between the current and the applied voltage in the circuit?

$$\tan\theta = I_{LC}/I_R = 1.2/0.44 = 2.73 \qquad \theta = 69.9°$$

Since I_C is larger than I_L, the current leads the applied voltage by 69.9°.

Problem 34. How much power is consumed in the circuit?

Power consumed is true power, since the values of the circuit resistance and current I_R are known, the true power can be calculated from the equation:

$$P_{TRUE} = I_R^2 R = (0.44)^2 \times 500 = 97 \text{ watts}$$

summary

☐ To solve for the impedance of a parallel LCR circuit, the parallel LC circuit is solved first, and then the net reactance is combined with the resistor as a parallel RL or RC circuit. The parallel LC circuit is characterized by the smaller of the two reactances. ☐ The voltage across a parallel LCR circuit is used as the phase reference.

☐ The currents in the three branches of a parallel LCR circuit are independent, and are found by: $I_L = E_{APP}/X_L$; $I_C = E_{APP}/X_C$; and $I_R = E_{APP}/R$. ☐ The two reactive currents are added as in a parallel LC circuit: $I_{LC} = I_L - I_C$, for I_L larger than I_C; and $I_{LC} = I_C - I_L$, for I_C larger than I_L. ☐ The line current is found by: $I_{LINE} = \sqrt{I_R^2 + I_{LC}^2}$. ☐ The phase angle is found by: $\tan\theta = I_{LC}/I_R$; or $\tan\theta = R/X$; or $\cos\theta = Z/R$. It is leading or lagging, depending on the relative magnitudes of the inductive and capacitive currents.

☐ The impedance of a parallel LCR circuit is found by finding the net reactance, X, of the L and C branches and combining it vectorially with R. The equation for the impedance is $Z = XR/(\sqrt{X^2 + R^2})$. ☐ The circuit will behave as an RL or RC circuit, depending on which reactance is the smaller. ☐ The impedance can also be found by $Z = E_{APP}/I_{LINE}$. ☐ The effect of frequency on a parallel LCR circuit is similar to that on parallel LC circuits, except that the maximum impedance can never be larger than the value of the resistance.

review questions

For Questions 1 to 5, consider a parallel LCR circuit with a resistance of 25 ohms, an inductive reactance of 50 ohms, and a capacitive reactance of 75 ohms.

1. What is the circuit impedance?
2. What is the net reactance of the circuit?
3. If the frequency is doubled, what is the impedance?
4. If the frequency is halved, what is the impedance?
5. If the capacitance is 2 microfarads, what is the inductance?

For Questions 6 to 8, consider a parallel LCR circuit with an applied voltage of 100 volts, and a line current of 20 amperes.

6. If the phase angle is 30 degrees, what is the resistance?
7. What is the power factor of the circuit?
8. What is the apparent power of the circuit?
9. For a parallel LCR circuit, the resistive, inductive, and capacitive currents are 10, 22.5, and 15 amperes, respectively. What is the line current?
10. What is the phase angle of the circuit of Question 9?

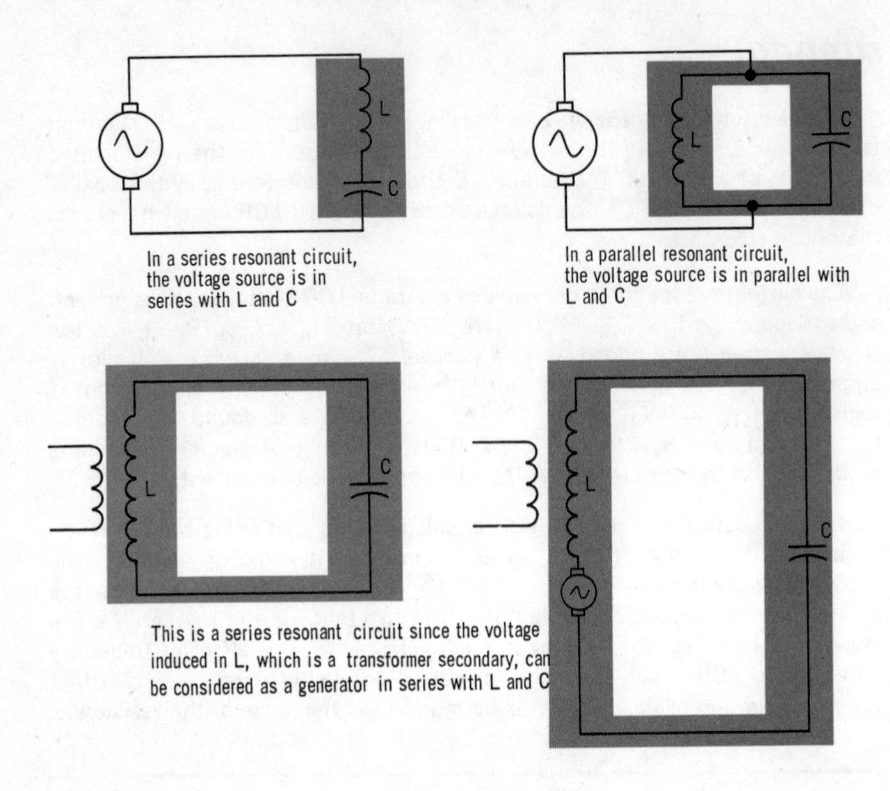

In a series resonant circuit, the voltage source is in series with L and C

In a parallel resonant circuit, the voltage source is in parallel with L and C

This is a series resonant circuit since the voltage induced in L, which is a transformer secondary, can be considered as a generator in series with L and C

parallel resonance

In parallel LC circuits, parallel resonance is the equivalent of series resonance in series LCR circuits. However, the *characteristics* of parallel resonance are quite *different* from those of series resonance.

For any given values of inductance and capacitance, the *frequency* at which parallel resonance takes place is *identical* to the frequency at which series resonance would take place for the same values of L and C. Therefore, parallel resonance can also be found by:

$$f_R = \frac{1}{2\pi\sqrt{LC}}$$

This being the case, the other equations you learned for finding L or C can also be used here:

$$L = \frac{1}{4\pi^2 f_R{}^2 C} \quad \text{and} \quad C = \frac{1}{4\pi^2 f_R{}^2 L}$$

Since it is sometimes difficult to distinguish series from parallel resonant circuits, think of a series resonant circuit as one with the voltage source in series with L and C; and a parallel resonant circuit as one with the voltage source in parallel with L and C.

the tank circuit

The properties of a parallel resonant circuit are based on the action that takes place between the parallel inductance and capacitance which is often called a *tank circuit,* because it has the ability to *store* electrical energy.

The action of a tank circuit is basically one of *interchange* of energy between the inductance and capacitance. If a voltage is *momentarily* applied across the tank circuit, C charges to this voltage. When the applied voltage is removed, C discharges through L, and a magnetic field is built up around L by the discharge current. When C has discharged, the field around L collapses, and in doing so induces a current that is in the same direction as the current that created the field. This current, therefore, charges C in the opposite direction. When the field around L has collapsed, C again discharges, but this time in the direction opposite to before. The discharge current again causes a magnetic field around L, which when it collapses, charges C in the same direction in which it was initially charged.

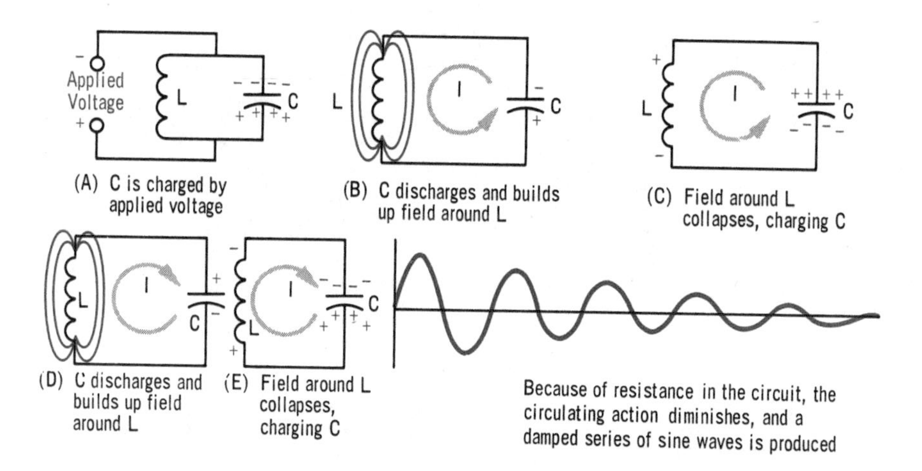

(A) C is charged by applied voltage

(B) C discharges and builds up field around L

(C) Field around L collapses, charging C

(D) C discharges and builds up field around L

(E) Field around L collapses, charging C

Because of resistance in the circuit, the circulating action diminishes, and a damped series of sine waves is produced

This interchange of energy, and the circulating current it produces, would continue *indefinitely* producing a series of *sine* waves, if we had an *ideal* tank circuit with no resistance. However, since some resistance is always present, the circulating current gradually diminishes as the resistance dissipates the energy in the circuit in the form of heat. This causes the sine-wave current to be *damped out.* If a voltage was again momentarily applied across the circuit, the interchange of energy and accompanying circulating current would begin again.

current and impedance at resonance

When an a-c voltage is applied to any parallel LC circuit having zero resistance, the currents in the inductive and capacitive branches are equal to:

$$I_L = E/X_L \qquad I_C = E/X_C$$

At resonance, X_L equals X_C, so the two currents are also equal. And since in a parallel LC circuit the two currents are 180 degrees out of phase, the line current, which is their vector sum, must be *zero*. Thus, the only current is the circulating current in the tank circuit. No line current flows. And if no line current flows, this means that the circuit has *infinite impedance* as far as the voltage source is concerned.

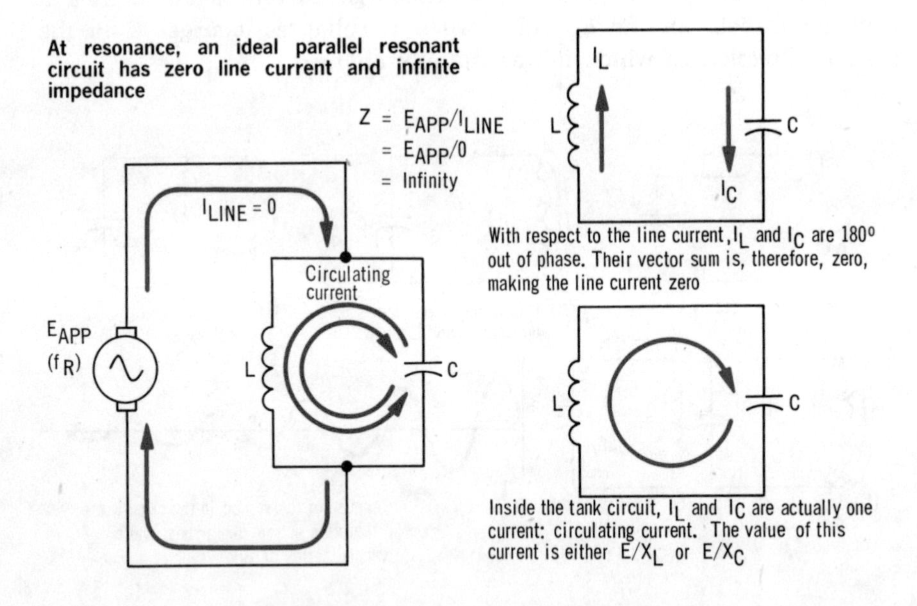

These two conditions of zero line current and infinite impedance are characteristic of *ideal* parallel resonant circuits at resonance. In practical circuits, which contain some resistance, the theoretical conditions of zero line current and infinite impedance are not realized. Instead, practical parallel resonant circuits have *minimum line current* and *maximum impedance* at resonance. You will recognize that this is exactly *opposite* to series resonant circuits, which have maximum current and minimum impedance at resonance.

current and impedance off resonance

In the ideal parallel resonant circuit at resonance, the branch currents, I_L and I_C, are equal, so the line current is zero and the circuit impedance is infinite. Above and below the resonant frequency, one of the reactances (X_L or X_C) is larger than the other. The two *branch currents* are therefore *unequal,* and the *line current,* which equals their vector sum (or arithmetic difference), has some value *greater than zero.* And since line current flows, the circuit *impedance is no longer infinite.* The further frequency is from the resonant frequency, the greater is the difference between the values of the reactances. As a result, the larger is the line current, and the smaller is the circuit impedance.

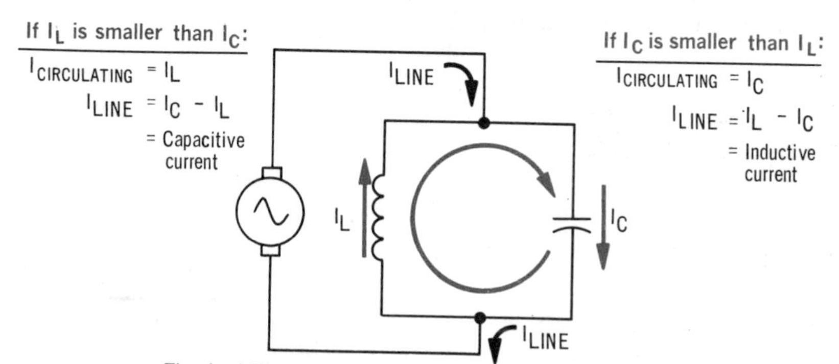

The circulating current in the tank circuit equals the smaller of the two branch currents. The line current is equal to the difference between the two branch currents, and has the same phase relationship as the larger of the two

The impedance at any frequency can be calculated from the equation you learned previously for the impedance of a parallel LC circuit. That is,

$$Z = \frac{X_L \times X_C}{X_C + X_L}$$

Where X_L and X_C are the reactances at the particular frequency involved. Also, as you learned for parallel LC circuits, the impedance is always greater than at least one of the reactances, instead of always being less than both, as it is for other types of parallel a-c circuits.

current and impedance off resonance (cont.)

At frequencies off resonance, the line current is equal to the difference (vector sum) between the values of the branch currents. Circulating current still flows in the tank circuit, and is equal to the *smaller* of the two branch currents. Thus, if I_L is 5 amperes and I_C is 3 amperes, the circulating current is 3 amperes and the line current is 2 amperes. In effect, the line current is that portion of the larger branch current that does not take part in the circulating current of the tank circuit. Since branch current I_L is inductive and branch current I_C is capacitive, the line current is *inductive* if I_L is *larger* (which means that X_L is smaller than X_C), and *capacitive* if I_C is *larger* (which means that X_C is smaller than X_L). This is the opposite of series resonant circuits, which you remember are inductive when X_L is larger, and capacitive when X_C is larger.

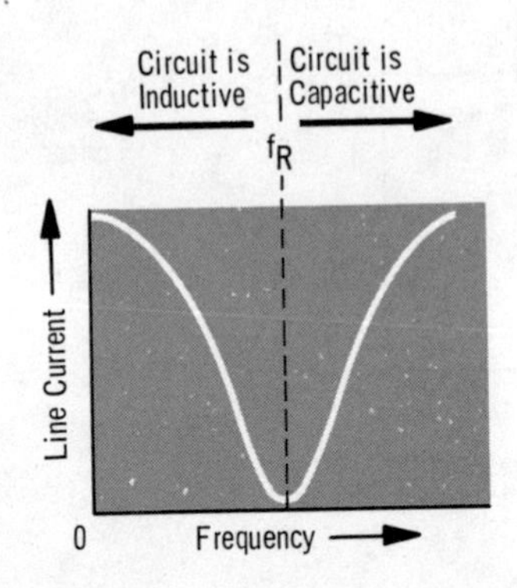

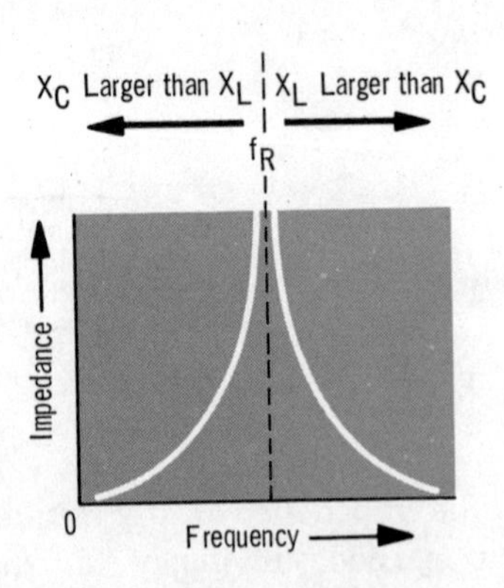

Curves showing how the line current and impedance of an ideal parallel resonant circuit change as the frequency is varied are shown. Notice that the characteristic shape of the *current*-vs.-frequency curve for the parallel resonant circuit is the same as the *impedance*-vs.-frequency curve for series resonant circuits. Similarly, the impedance-vs.-frequency curve for a parallel circuit is the same as the current-vs.-frequency curve for series circuits.

practical parallel resonant circuits

Practical parallel resonant circuits differ from the ideal parallel circuit just described in one major respect: practical circuits contain *resistance*. This resistance is contained in the inductance, the capacitance, and the interconnecting wires. Normally, however, only the resistance of the inductance is large enough to be considered. In analyzing a circuit, this resistance is considered to be in *series* with the *inductance*. A practical parallel resonant circuit, therefore, consists of a purely capacitive branch, and an inductive branch that is actually a series RL circuit. You will remember from what you have learned about the 10:1 ratio, that if the inductive reactance of the inductance is 10 times or more greater than its resistance, the resistance can generally be neglected. The circuit could then be analyzed in the same manner as the ideal parallel resonant circuit. For purposes of explanation, though, we will include the effects of the resistance regardless of its value relative to X_L.

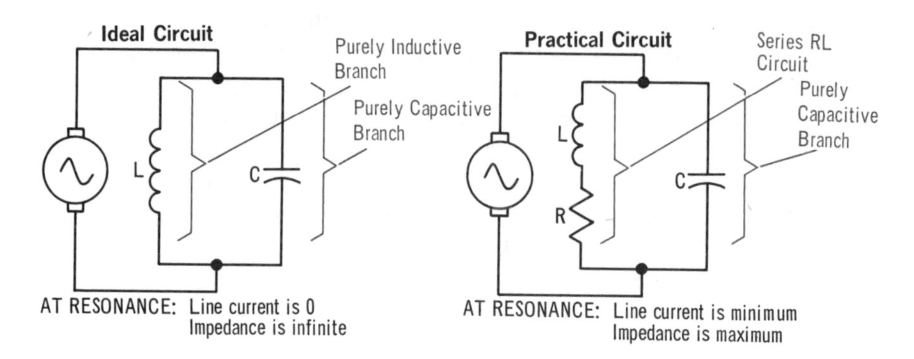

The principal effect of the resistance in a parallel resonant circuit is that it causes the current in the inductive branch to lag the applied voltage by a phase angle of *less* than 90 degrees, instead of exactly 90 degrees as is the case in the ideal circuit. As a result, the two branch currents are not 180 degrees out of phase. For simplicity, resonance can still be considered as occurring when X_L equals X_C, but now when the two branch currents are added vectorially, their sum is *not* zero. This means that at resonance some line current flows. And since there is line current, the impedance cannot be infinite, as it is in the ideal circuit. Thus, at resonance, practical parallel resonant circuits have *minimum* line current and *maximum* resistance, instead of zero line current and infinite impedance, as do ideal circuits.

practical parallel resonant circuits (cont.)

At resonance, the *line current* in a practical parallel resonant circuit is *in phase* with the *applied voltage*. The reason for this is that since X_L and X_C are equal, the reactive (inductive) component of the current in the RL branch is equal to, and cancels, the current (capacitive) in the capacitive branch. Only the *resistive* component of the current in the RL branch, therefore, flows in the line. Since the capacitive branch contains no resistance, the current in it is equal to:

$$I_C = E/X_C$$

The current in the inductive branch is calculated the same as in any series RL circuit, and so is equal to:

$$I_L = E/\sqrt{R^2 + X_L^2}$$

The line current can then be found by adding the two branch currents *vectorially*. However, the branch currents differ in phase by less than 180 degrees, but more than 90 degrees. Therefore, to add them vectorially you cannot use the Pythagorean Theorem, since it applies only to quantities 90 degrees apart, and you cannot subtract them arithmetically, since this only applies to quantities 180 degrees apart. They can, however, be added by first being resolved into their vertical and horizontal components, then adding the components, and finally finding the resultant of the total components. Addition of vectors by components was described earlier. At resonance, the calculation of the line current, as well as the circulating current in the tank and the circuit impedance, can be done much more easily by using the Q of the circuit.

Branch currents are added vectorially by components to find the line current in practical parallel resonant circuits

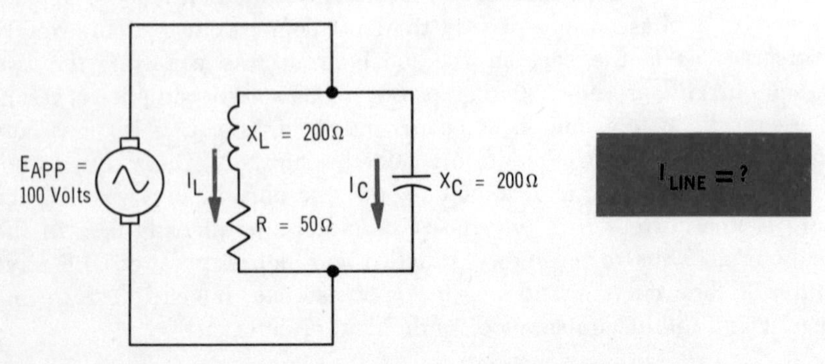

practical parallel resonant circuits (cont.)

The branch currents, I_L and I_C, of the circuit on page 4-138 are

$$I_L = \frac{E}{\sqrt{R^2 + X_L^2}} = \frac{100}{\sqrt{(50)^2 + (200)^2}} = 0.486 \text{ ampere}$$

$$\tan\theta = X_L/R = 200/50 = 4 \qquad \theta = 76°\text{; current lagging voltage}$$

$$I_C = E/X_C = 100/200 = 0.5 \text{ ampere}$$

The horizontal (H) and vertical (V) components of the branch currents are

$$I_{L(H)} = I_L \cos\theta = 0.486 \times 0.242 = 0.118 \text{ ampere}$$

$$I_{L(V)} = I_L \sin\theta = 0.486 \times 0.97 = 0.47 \text{ ampere}$$

$$I_{C(H)} = 0\text{; no resistive component} \qquad I_{C(V)} = 0.5 \text{ ampere}$$

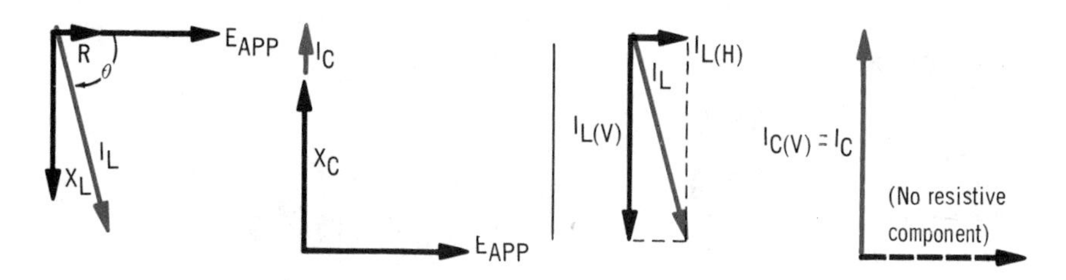

The total horizontal and vertical components are

$$I_{TOT(H)} = I_{L(H)} + I_{C(H)} = 0.118 + 0 = 0.118 \text{ ampere}$$

$$I_{TOT(V)} = I_{C(V)} - I_{L(V)} = 0.5 - 0.47 = 0.03 \text{ ampere}$$

The minus sign of equation $I_{TOT(V)}$ indicates that $I_{L(V)}$ vector points down, or lags the voltage.

The resultant of the total components, which is the line current, is

$$I_{LINE} = \sqrt{I_{TOT(H)}^2 + I_{TOT(V)}^2} = \sqrt{(0.118)^2 + (0.03)^2} = 0.123 \text{ ampere}$$

The phase angle between the line current and the applied voltage is

$$\tan\theta = \frac{I_{TOT(V)}}{I_{TOT(H)}} = \frac{0.03}{0.118} = 0.254$$

$$\theta = 14.3°\text{; current leading voltage}$$

the resonance band

You will remember that for every series resonant circuit there is a *range* of frequencies above and below the resonant frequency at which, for practical purposes, the circuit can be considered as being at resonance. This range of frequencies was called the *resonance band* or *bandpass,* and consisted of all the frequencies at which the circuit current was 0.707 or more times its value at resonance. Parallel resonant circuits also have a resonance band; but it is defined in terms of the *impedance-vs.-frequency curve,* and consists of all the frequencies that produce a circuit impedance 0.707 or more times the impedance at resonance.

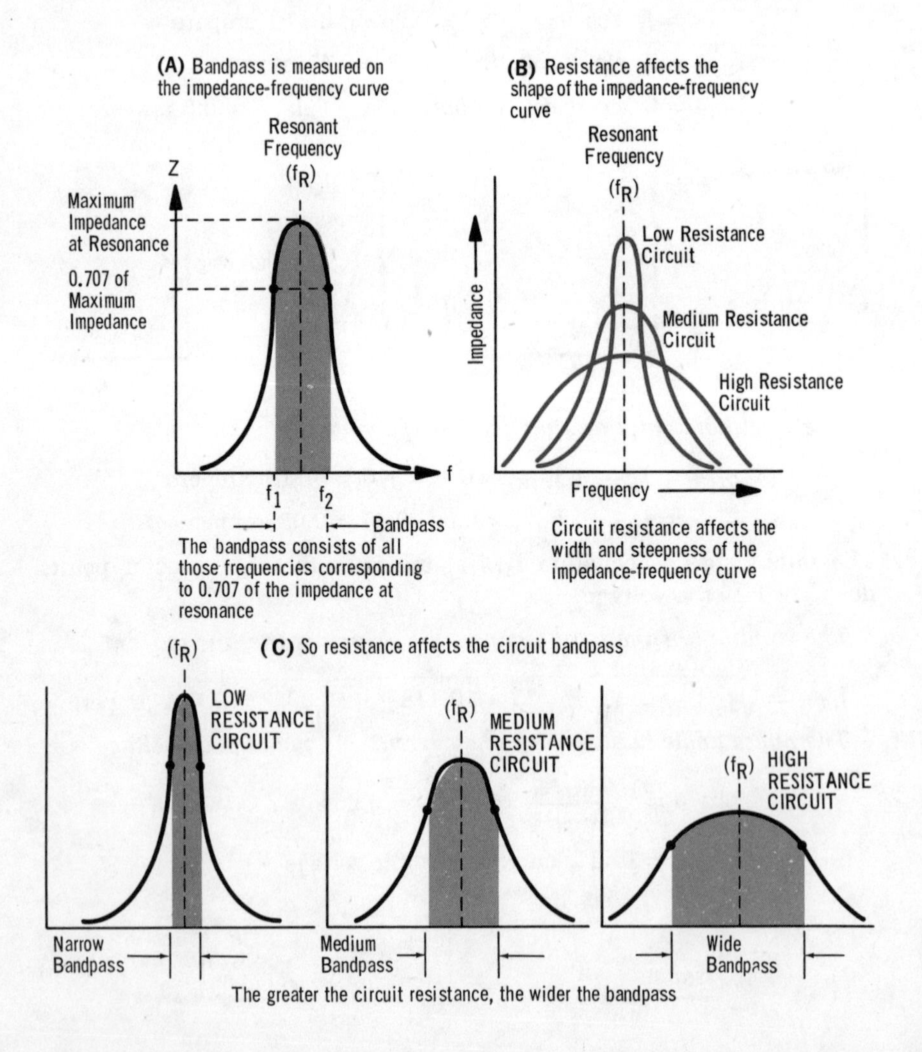

the Q of a parallel resonant circuit

The Q, or quality, of a series resonant circuit, you will recall, is determined by the ratio of the voltage across either X_L or X_C to the applied voltage. For parallel resonance, the Q also measures the quality of a circuit. However, in parallel resonant circuits, Q is not determined on the basis of voltage, but rather on the basis of current. The Q of a parallel resonant circuit is defined as the *ratio* of the current in the *tank* to the *line* current. Thus,

$$Q = I_{TANK}/I_{LINE}$$

where I_{TANK} is the circulating current in the tank. Mathematically, this equation can be converted to the form:

$$Q = X_L/R$$

A High Q Means a Narrow Bandpass

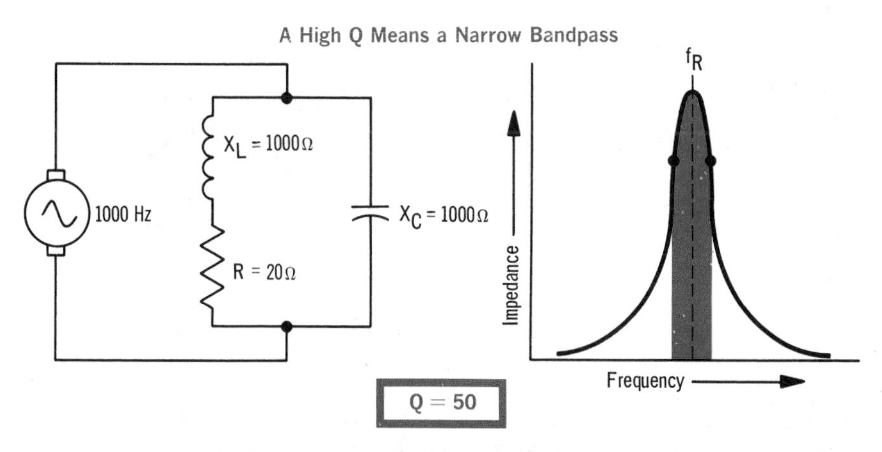

The Q of a parallel resonant circuit is its ability to discriminate between resonant and nonresonant frequencies. The Q is the ratio of I_{TANK} to I_{LINE}, but can be calculated more easily from the equation $Q = X_L/R$. In terms of its Q, the bandpass of a parallel resonant circuit is Bandpass $= f_R/Q$.

You will recognize this as the same equation used for the Q of a series resonant circuit. As a result, resistance has the same effect on the Q of a parallel resonant circuit as it does on a series resonant circuit. The *lower* the *resistance*, the *higher* is the Q of the circuit, and the narrower is its bandpass. Conversely, the *greater* the *resistance*, the *lower* is the Q, and the *wider* is the *bandpass*.

the Q of a parallel resonant circuit (cont.)

If the Q of a circuit is known, the bandpass can be calculated by the equation:

$$\text{Bandpass (Hz)} = f_R/Q$$

Circuits with high Q's, therefore, can discriminate between resonant and nonresonant frequencies better than circuits with low Q's. In practical applications, when the Q of a parallel resonant circuit is 10 or greater, this resistance can be neglected, and the circuit considered as an almost ideal parallel resonant circuit.

A Low Q Means a Wide Bandpass

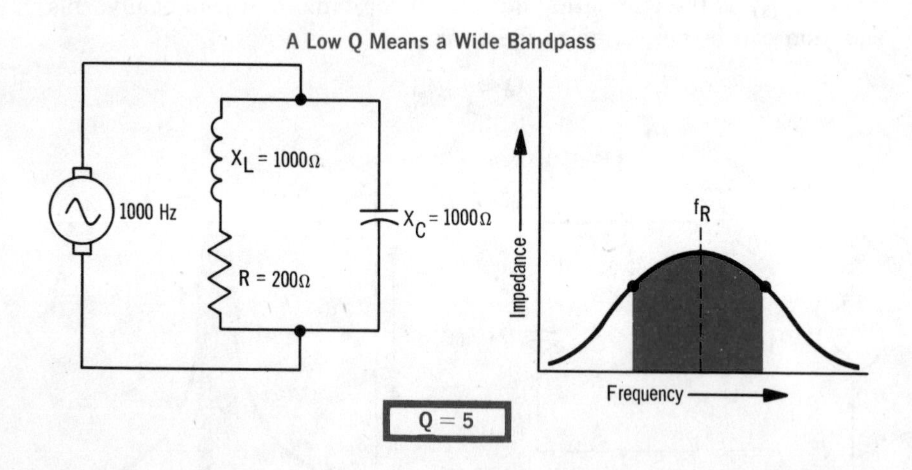

Once the Q is known, the values at resonance of the other important circuit quantities can easily be found. These quantities are the impedance (Z), the tank current (I_{TANK}), and the line current (I_{LINE}). The equations for impedance in terms of Q are

$$Z = QX_L$$

or

$$Z = QX_C$$

The equation for I_{TANK} in terms of Q is

$$I_{TANK} = QI_{LINE}$$

The equations for I_{LINE} in terms of Q are

$$I_{LINE} = E/QX_L$$

or

$$I_{LINE} = E/QX_C$$

controlling Q

For the most part, resonant circuits are desiged to have high Q's, so the circuits can be most efficient. Also the Q that results usually depends on the inherent design of the coil, i.e., the internal resistance of the coil that develops from the way the coil is wound.

However, in some instances there is a need for the Q to be a specific value, which may be lower than the Q that would ordinarily result from a well-designed tank circuit. A good example of this is when a specific bandpass might be wanted at the resonant frequency, and this specific bandpass would require a specific Q.

Since the basic equation for Q is $Q = X_L/R$, one method of controlling Q would be to make R the specific value needed. For example, if X_L is 1000 ohms, and you wanted a Q of 5, then you would have to make sure that R would be 200 ohms. This would require you to add a resistor in series with the inductor that, when added to the a-c resistance of the windings, would produce a total resistance of 200 ohms. Q, then, would be:

$$Q = \frac{X_L}{R_L + R_{series}}$$

This would also be true for series resonant circuits.

Another way to control the Q, which is easier in parallel resonant circuits, is to use a shunt resistor across the tank. Since Q in a parallel resonant circuit depends on I_{TANK}/I_{LINE}, a shunt resistor across the tank would increase I_{LINE} to lower the Q. The value of I_{LINE}, of course, depends on tank impedance, Z_{TANK}. By manipulating the equation, $Z = QX_L$, which you studied earlier, you can see that $Q = Z/X_L$. By adding R_{SHUNT} to the circuit, the impedance Z will change, and the equation becomes:

$$Q = \frac{\left(\dfrac{Z_{TANK} \times R_{SHUNT}}{Z_{TANK} + R_{SHUNT}}\right)}{X_L}$$

which reduces to

$$Q = \frac{Z_{TANK} \times R_{SHUNT}}{X_L(Z_{TANK} + R_{SHUNT})}$$

$$Q = \frac{X_L}{R_L + R_{SERIES}}$$

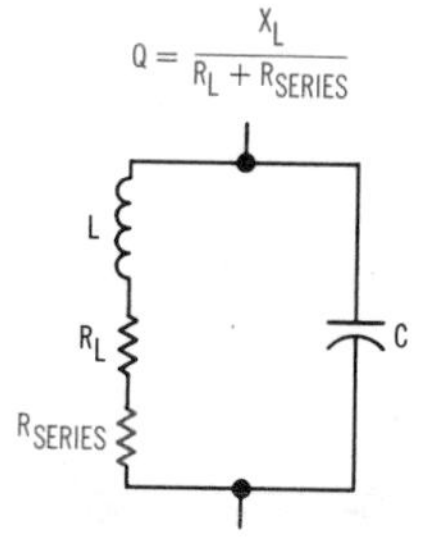

$$Q = \frac{I_{TANK}}{I_{LINE}}$$

The Q of a tank circuit, and therefore its bandpass, can be controlled by placing the proper value resistor in the choke leg or another value resistor across the tank. R_{SERIES} is usually a low value, and R_{SHUNT} is usually a high value in comparison. The shunt method is the most popular

$$Q = \frac{Z \times R_{SHUNT}}{X_L(Z + R_{SHUNT})}$$

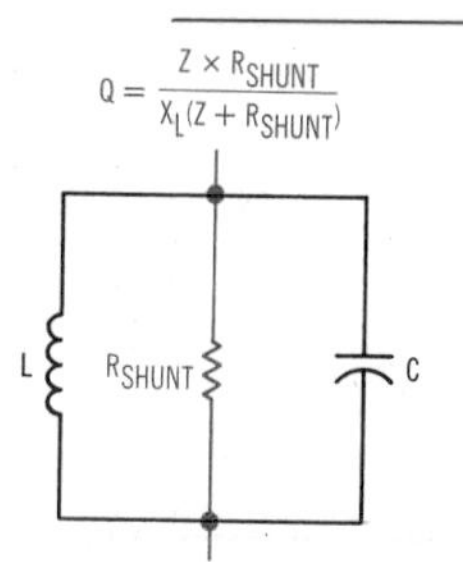

solved problems

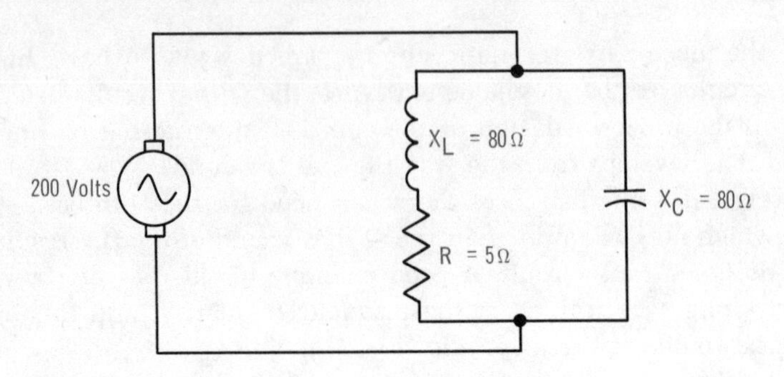

Problem 35. Is this circuit at resonance?

Remember that here X_L and X_C are equal only at resonant frequency. Since X_L and X_C both equal 80 ohms, the circuit must be at resonance.

Problem 36. What is the value of the current flowing in the tank?

One way to solve this problem is to use the equation $I_{TANK} = QI_{LINE}$. Of course Q and I_{LINE} must be determined first:

Calculating Q:

$$Q = X_L/R = 80/5 = 16$$

Calculating I_{LINE}:

$$I_{LINE} = E/Z = E/QX_L = 200/(16 \times 80) = 0.156 \text{ ampere}$$

Calculating I_{TANK}:

$$I_{TANK} = QI_{LINE} = 16 \times 0.156 = 2.5 \text{ amperes}$$

Thus, with only 0.156 ampere in the line, 2.5 amperes flow in the tank.

Problem 37. If the resonant frequency is 2000 Hz, what is the circuit bandwidth?

Using the Q of the circuit, calculate bandwidth from:

$$\text{Bandwidth (Hz)} = f_R/Q = 2000/16 = 125 \text{ hertz}$$

The bandwidth, therefore, extends from about 1938 Hz (2000 − 62.5) to about 2063 Hz (2000 + 62.5).

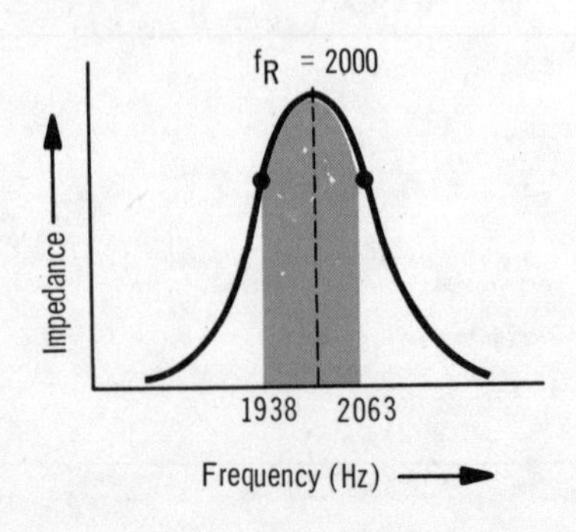

comparison of series and parallel resonant circuits

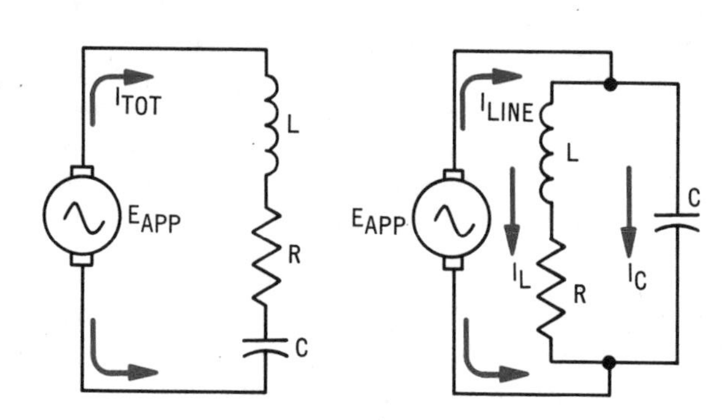

PROPERTIES AT RESONANCE

	Series Resonant Circuit	Parallel Resonant Circuit
Resonant Frequency (f_R)	$\frac{1}{2\pi\sqrt{LC}}$	$\frac{1}{2\pi\sqrt{LC}}$
Reactances	$X_L = X_C$	$X_L = X_C$
Impedance	Minimum; $Z = R$	Maximum; $Z = QX_L$
Current (I_{TOT} or I_{LINE})	Maximum; I_{TOT}	Minimum; I_{LINE}
Q, Quality	$E_L/E_{APP} = X_L/R$	$I_{TANK}/I_{LINE} = X_L/R$
Bandwidth	f_R/Q	f_R/Q

PROPERTIES OFF RESONANCE

	Series Resonant Circuit		Parallel Resonant Circuit	
	Above f_R	Below f_R	Above f_R	Below f_R
Reactances	$X_L > X_C$	$X_C > X_L$	$X_L > X_C$	$X_C > X_L$
Impedance	Increases	Increases	Decreases	Decreases
Phase Angle Between E_{APP} and I_{TOT} or I_{LINE}	I lags E	I leads E	I leads E	I lags E
Inductive or Capacitive Circuit	Inductive	Capacitive	Capacitive	Inductive

summary

☐ The properties of a parallel resonant circuit are based on the interchange of energy between an inductor and capacitor connected in parallel across a voltage source. Such a circuit is often called a tank circuit. ☐ For a given inductor and capacitor, the frequency at which parallel resonance occurs can be found by: $f_R = 1/(2\pi\sqrt{LC})$. ☐ Theoretically, a parallel resonant circuit has infinite impedance and zero line current. Practical parallel resonant circuits, however, exhibit maximum impedance and minimum line current at resonance. ☐ For frequencies above or below the resonant frequency, the line current of a parallel resonant circuit increases progressively, while its impedance decreases.

☐ Practical parallel resonant circuits contain resistance, as well as inductance and capacitance. Normally this resistance is the resistance of the inductor wire, and can be considered as being in series with the inductance. ☐ The principal effect of the resistance is that at resonance the two branch currents are not 180 degrees out of phase, so line current flows in the circuit. ☐ The magnitudes of the branch currents can be found by: $I_C = E/X_C$ and $I_L = E/(\sqrt{R^2 + X_L^2})$. ☐ The line current can be determined by adding the line currents vectorially.

☐ The bandpass of a parallel resonant circuit consists of all frequencies that produce a circuit impedance 0.707, or more, times the impedance at resonance. ☐ The bandpass can be found from the Q of a circuit by: Bandpass (cps) $= f_R/Q$. ☐ The Q, or quality, of the circuit is the ratio of the circulating current in the tank (I_{TANK}) to the line current (I_{LINE}), or: $Q = I_{TANK}/I_{LINE}$.

review questions

1. What is an *ideal parallel resonant circuit*?
2. How does a practical parallel resonant circuit differ from an ideal one?
3. What is the resonant frequency of a 50-microfarad capacitor and 50-millihenry coil connected in parallel?
4. How is the circulating current produced in a tank circuit?
5. What is meant by *damping* in a tank circuit?
6. Draw an impedance-vs.-frequency curve for a parallel resoant circuit.
7. How is *bandwidth* of a parallel resonant circuit defined?
8. What is meant by the Q of a parallel resonant circuit?
9. How is Q related to resistance? To bandwidth?
10. If a parallel resonant circuit has a frequency of 1000 Hz and a Q of 10, what is its bandwidth?

tuning

Until now, you studied resonant circuits in which fixed values of L and C were used to produce the desired resonant frequency. In actual practice this is impractical, because coils and capacitors cannot be made precisely enough to get the exact tuned frequency. These parts have tolerance values, and rarely do any two parts have the exact same value. Also, quite often, it is important to have the resonant circuit function at different frequencies, or even over different ranges of frequencies. To do this, either the inductor or capacitor or both must be variable, so that they can be tuned for the precise frequency; and when tuning ranges are required, different inductors and/or capacitors can be switched in or out of the tank circuit.

The simplest form of tuning can be used with a tank circuit that is designed to work at only one resonant frequency. The tuning adjustment, then, is needed only to compensate for the tolerance variations of the parts. In these cases, the basic L and C components can be fixed, and a small *trimmer capacitor* can be added to make minor *fine tuning* adjustments.

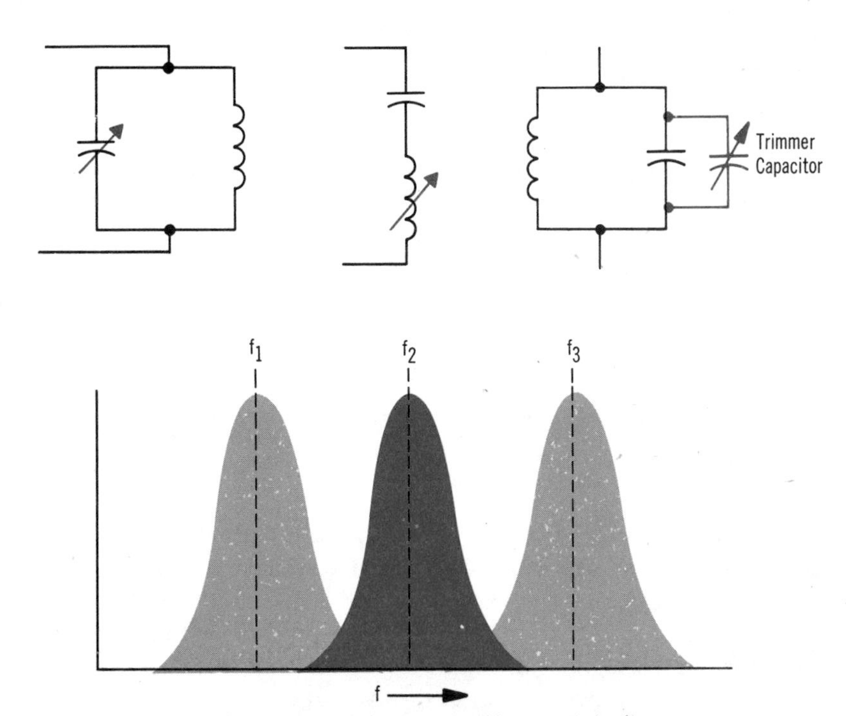

If variable capacitors or inductors are used in resonant circuits, the resonance point and bandpass frequencies can be changed to a variety of frequencies by a simple adjustment.

tuning (cont.)

For *broadband tuning*, either the inductor or capacitor should be variable. Generally speaking, there are more standard variable capacitors available than variable inductors, so it is usually easier to design a tunable circuit with a variable capacitor. Also, it is easier to design a variable capacitor to give a wider range of control than a variable inductor; so *broadband resonant circuits* are more often capacitor tuned. Many variable capacitors used in radios have built-in trimmer capacitors to adjust the minimum-to-maximum tracking of the capacitor and, thus, tuning variability.

This does not mean that variable inductors are not used for broadband tuning. They are, particularly since they allow delicate, small adjustments to be made more easily. Since variable inductors use plungers that move in and out of the coil, complicated mechanical arrangement is needed when tuning is made frequently, usually from the front panel of a piece of equipment. A variable capacitor does that job much more simply. But for tuning adjustments that are made infrequently, the variable inductor will work as well as or better than the variable capacitor. With high frequencies, the tank circuit may not have an actual capacitor across the coil. Instead, the innerwiring capacities of the variable inductor resonate with the coil.

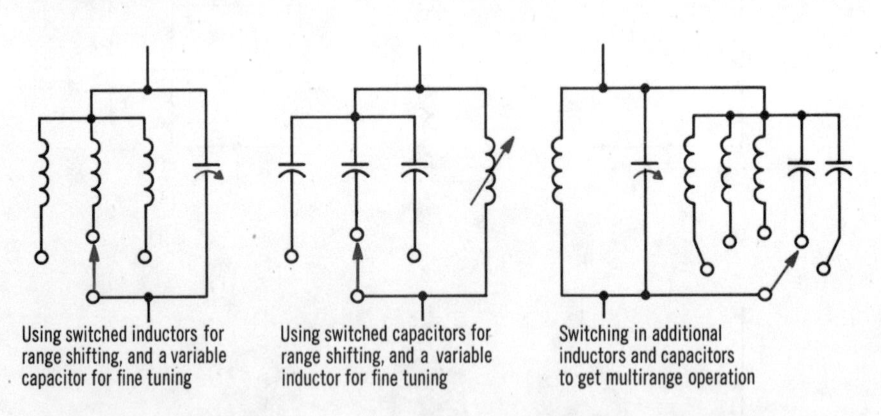

Using switched inductors for range shifting, and a variable capacitor for fine tuning

Using switched capacitors for range shifting, and a variable inductor for fine tuning

Switching in additional inductors and capacitors to get multirange operation

Sometimes, both the coil and capacitor are variable, one acting as a coarse adjustment, and the other as a fine or vernier adjustment. In some circuits, the range of frequencies that the resonant circuit must work through is so broad that no one variable capacitor or inductor can provide enough variability. In these cases, either different inductors or capacitors can be switched in and out of the tuned circuits to provide a series of frequency ranges, or additional components can be added to those already in the circuit to shift the range of operation constantly.

filter circuits

As used in electrical circuits, the term filter means to offer large opposition to, or *reject*, voltages and currents of certain frequencies, and at the same time offer little opposition to, or *pass*, voltages and currents of other frequencies. Circuits which have this capability are called *filter circuits*. Filter circuits are divided into various groups, according to the frequencies they pass and reject. One such group passes all frequencies up to a certain frequency, and rejects all above that frequency. This type of filter is called a *low-pass filter*. Another group of filters rejects all frequencies up to a certain frequency, and passes all those higher than that frequency. These are called *high-pass filters*. Other filters pass certain ranges, or bands, of frequencies and reject all frequencies outside of the band. Such filters are called *band-pass* filters. Still others reject a band of frequencies, and pass all frequencies outside of the band. They are *band-reject* filters. These, then, are the principal types of filter circuits.

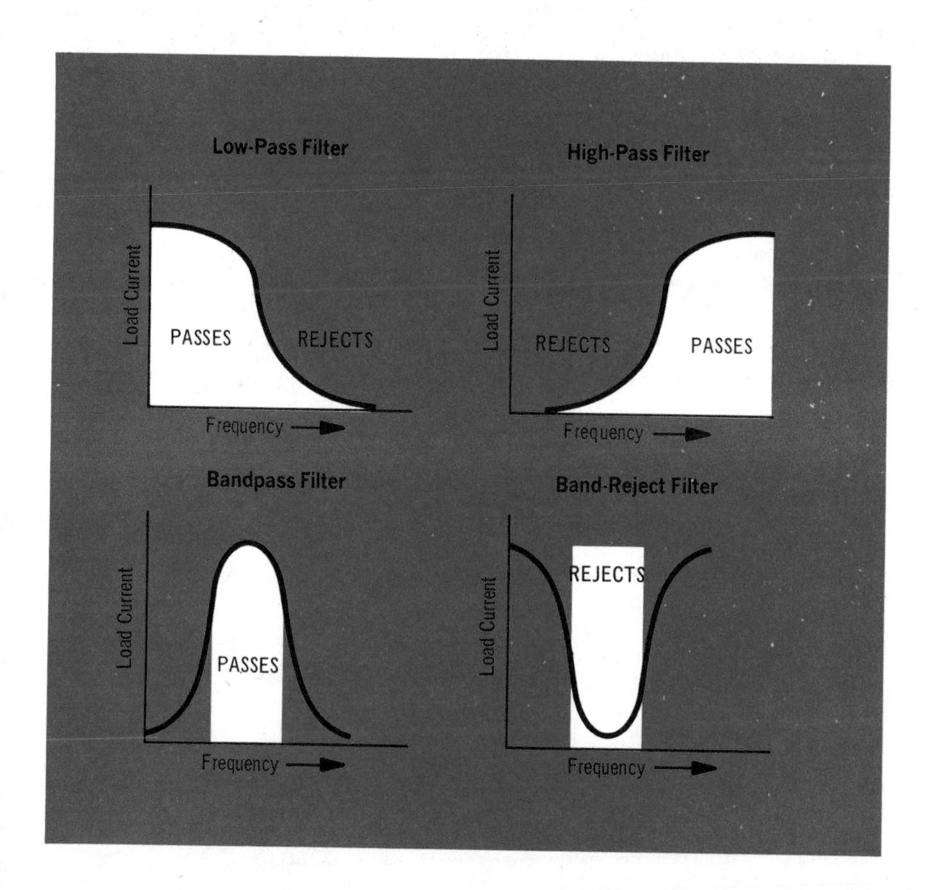

low-pass filters

The *low-pass filter* relies on the varying impedance of an inductor and/or capacitor to interfere with a signal as its frequency increases. In its simplest form a low-pass filter can be merely an inductor in series with the output load. At low frequencies, the reactance, X_L, of the inductor is small, so that very little signal is dropped across the inductor, leaving most of the signal to reach the output load. As the frequency increases, X_L increases; and when X_L reaches a value that is 1/10 of R, the drop across L will reduce the signal drop across R. The more the frequency rises, the greater the drop across L and the less the drop across R. At the frequency where X_L is equal to R, half the signal will be lost at the output. At and beyond the frequency where X_L is ten or more times greater than R, for all practical purposes, all of the signal will drop across L, leaving no signal output across R.

When a capacitor is used as a low-pass filter, it must be used differently since its reactance is inversely proportional to frequency. The capacitor is connected in parallel with the output load to bypass the signal. At low frequencies, though, its reactance, X_C, is high compared to the value of R, and so it has little effect. Its effect becomes noticeable, though, at the frequency where X_C decreases to the point at which it is about 10 times R. From there on, as the frequency is increased, more and more of the signal is bypassed around the load. At the frequency where X_L reduces to about 1/10 of R, effectively all of the signal will be shunted around the output.

When the inductor and capacitor are combined, their effects reinforce each other so that at or about the critical cutoff frequencies the passband curve drops more sharply and there is more effective filtering at the rejection frequencies. The design, though, differs because the effects of resonance must be considered. When $X_L = X_C$, the signal output peaks, after which the sharp drop follows.

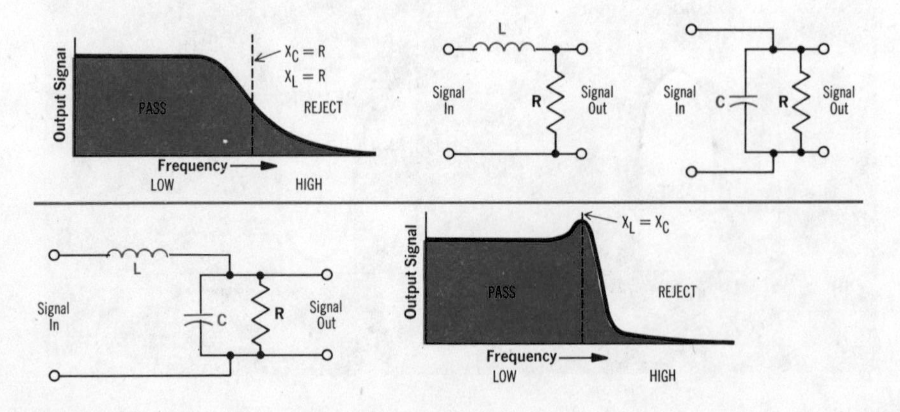

high-pass filters

The *high-pass filter* works in a manner similar to that of the low-pass filter in that it also relies on the varying reactance characteristics of the inductor and capacitor. However, since the high-pass filter is designed to block the lower frequencies, the parts are connected into the circuit differently. The inductor, for example, is *shunted* across the output load. At the lower frequencies, X_L will be very low compared to the load resistance, and will effectively short out the load. As the frequency increases and X_L rises, the shunting effect of L is proportionately reduced. For all practical purposes, when X_L reaches about 1/10 the value of R, some output signal starts to develop; and as X_L rises further with frequency, the output continues to climb until X_L is 10 or more times greater than R. Then the circuit works as though the inductor did not exist.

A capacitor would have to be connected in *series* with the output. Its reactance, X_C, would be very high at low frequencies, and so most of the signal would be dropped across C, leaving little left for the output. But with the higher frequencies, as X_C decreases to where it is only ten times the value of R, some signal would start developing at the output, and would rise with frequency as X_C continued to decrease until X_C reached 1/10 the value of R.

When L and C are combined in a high-pass filter, they are placed with C in series with the output, and L shunting the output. As with the low-pass filter, they sharpen the rejection slope and peak at resonance.

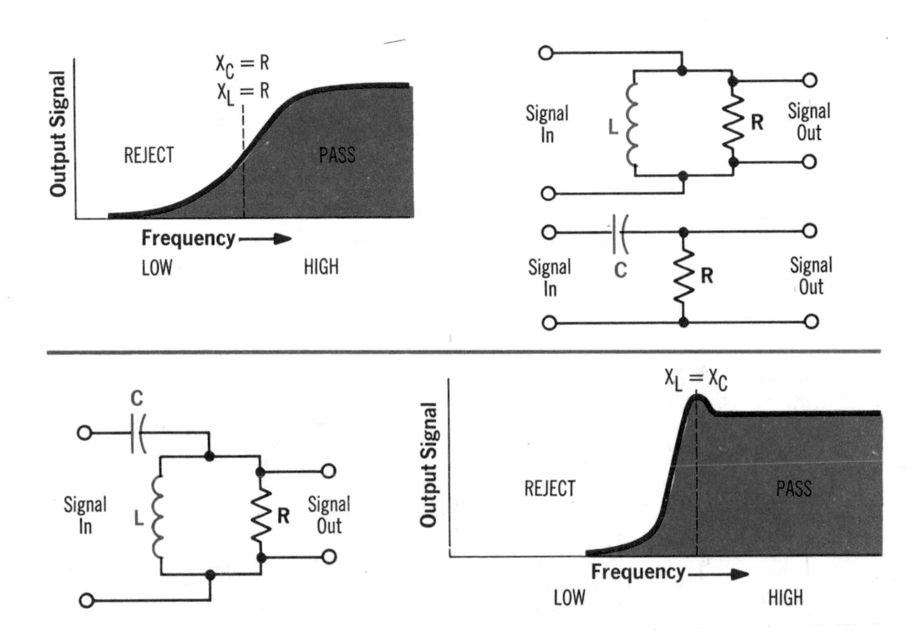

bandpass and band-reject filters

Whereas low- and high-pass filters work either above or below a certain point, the *bandpass filter* functions differently within a specific *range* of frequencies from the way it does at frequencies either above or below that range.

The ordinary resonant tank circuit in the output is the best example of a typical bandpass filter. The resonant frequency, f_R, of the tank will establish the center point of the band to be passed. At resonance, the tank will present a high load impedance at the output. But off resonance, the impedance will drop, offering less and less output load for a signal to develop. The width of the bandpass will depend on the Q of the tank; by controlling the Q, either a narrow or wide bandpass can be obtained.

The series-resonant circuit can function as a bandpass filter when it is connected in series with the output. The series resonant circuit has a very low impedance at resonance, and so allows most of the signal to reach the output. When above and below resonance, though, the high impedance of the series circuit (X_C at the lower frequencies and X_L at the higher frequencies) drops the signals before they reach the output.

Band-reject filters function in a manner exactly opposite that of bandpass filters. The resonant tank circuit placed in series with the output will use its high impedance at resonance to prevent the signal from reaching the output. And the series-resonant circuit shunted across the output will use its low impedance at resonance to bypass the output. In either case, frequencies above or below the resonant band will be onlv slightly affected.

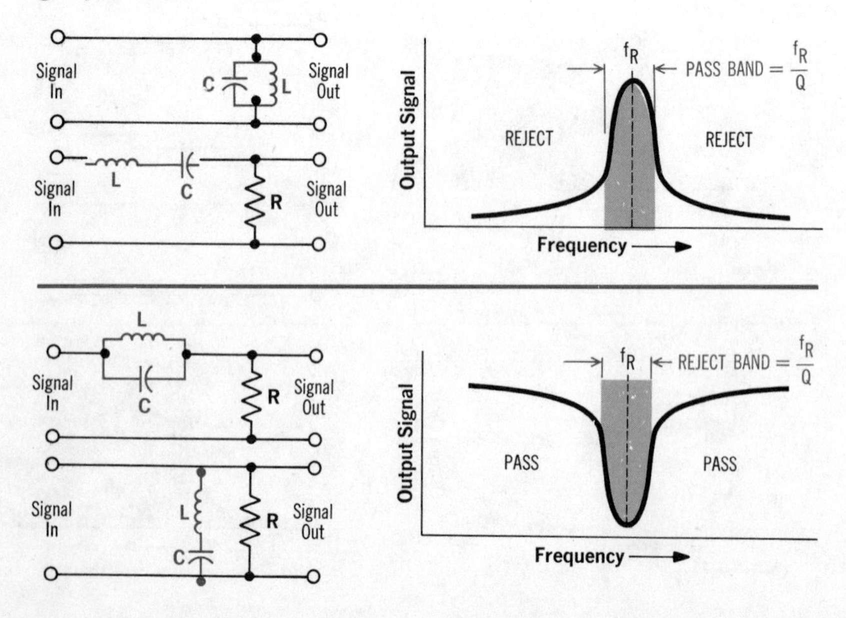

filter networks

Aside from the simple LC combinations shown on the previous pages, inductors and capacitors can be combined in a variety of ways to form many simple or intricate *filter networks*. The basic circuit patterns can be repeated to attain as high a degree of filtering as necessary.

Simple LC low- and high-pass filters are actually called *half-section filters* because they represent only a part of the more sophisticated filter circuits that are in use today. Typical filter sections generally form T or pi networks, so named because of the shape they take schematically. A low-pass *T filter* has two inductors in series with the signal, and a capacitor shunting the line from between the two inductors. A high-pass T filter has two capacitors in series with the signal line, and an inductor shunting the line from between the two capacitors. These T networks function in the same manner as *half-section filters*, except that the extra component will give added attenuation of the unwanted frequencies.

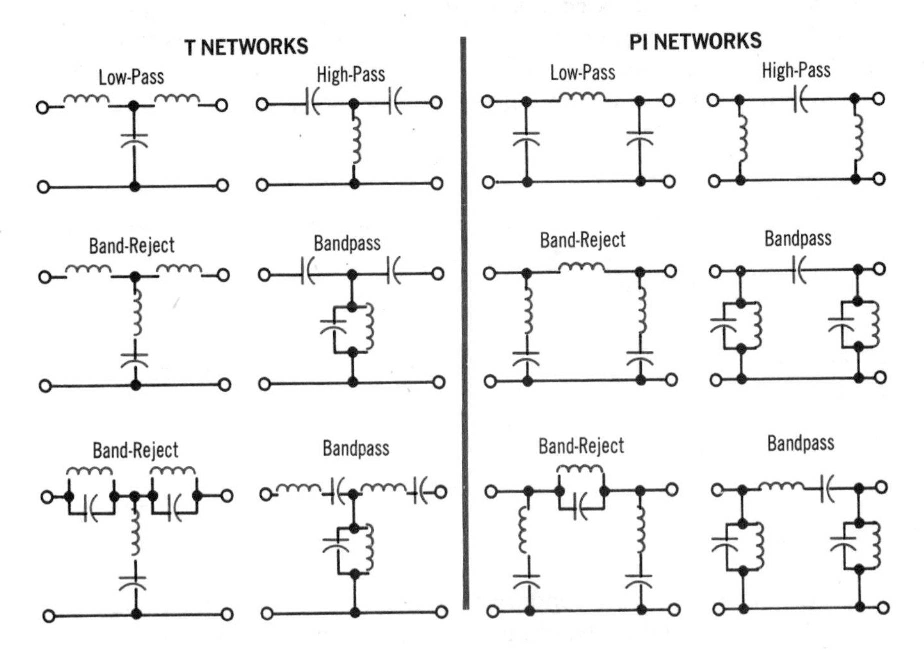

The *pi networks* are so called because they resemble the Greek letter pi (π). They use two shunt components, one on either side of the series component, and like the T filter, give better attenuation of unwanted frequencies.

For bandpass and band-reject networks, the individual inductors and/or capacitors can be replaced with resonant circuits to form a variety of T and pi networks.

transformer vectors

You recall that the theory of operation of ordinary transformers was given in Volume 3. Transformers were described then strictly on the basis of the current and voltage waveforms that exist in the primary and secondary. Transformer voltages and currents can also be represented by *vectors,* as shown below. When this is done, the *primary voltage* is used as the zero-degree *phase reference,* and the other voltage and current vectors are shown in relation to it.

In a transformer with a very small secondary current flowing, the primary current lags the primary voltage by 90 degrees, and the secondary current lags the secondary voltage by 90 degrees. Furthermore, the primary and secondary voltages are 180 degrees out of phase. On a vector diagram, therefore, the four vectors representing the currents and voltages are *mutually perpendicular,* as shown. As the secondary current increases, assuming that the secondary load is resistive, the current in the secondary begins to go resistive. As a result, it no longer lags the secondary voltage by 90 degrees. The more current that flows in the secondary, the more resistive it becomes, and the smaller the angle between the secondary voltage and current becomes. You will re-

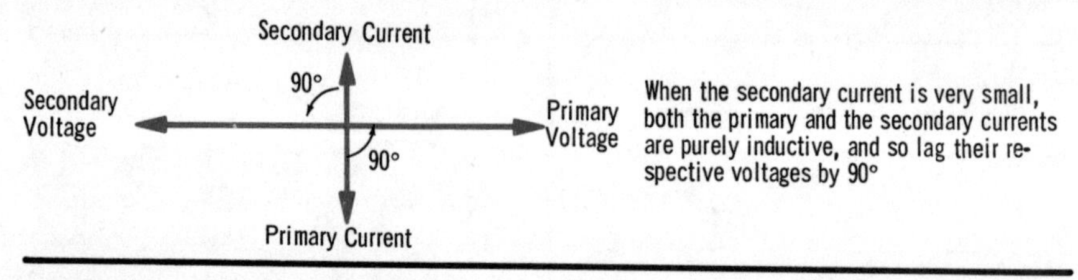

When the secondary current is very small, both the primary and the secondary currents are purely inductive, and so lag their respective voltages by 90°

When the rated current flows in the secondary, both the primary and secondary currents are basically resistive. As a result, they lag their respective voltages by only a small phase angle

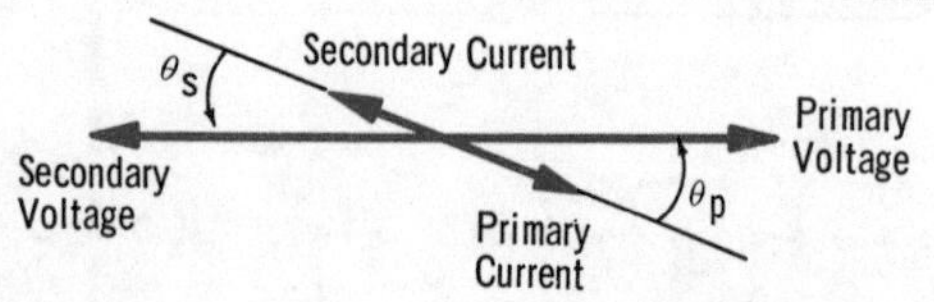

member that when the secondary current increases, the magnetic field created in the secondary winding causes the primary current to also increase by effectively reducing the inductive reactance of the primary winding. With the primary inductive reactance reduced, the primary also becomes more resistive, and so the angle between its current and voltage also decreases. Thus, the vector diagram of a transformer carrying its rated current shows that the primary and secondary voltages are 180 degrees out of phase, with the two currents lagging their respective voltages by the same small angle.

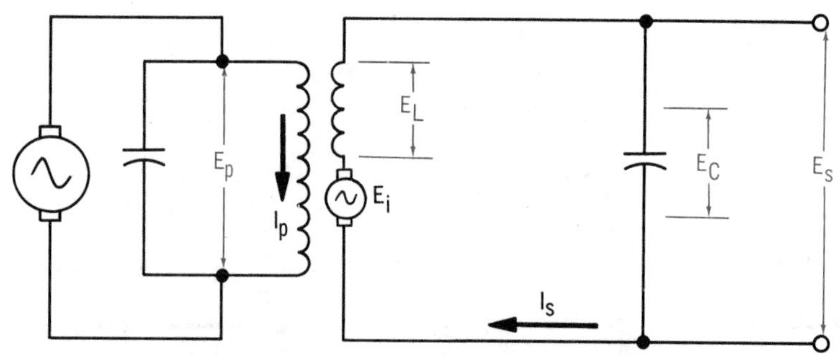

In tuned transformers, the output, or secondary voltage (E_S) is not actually the voltage induced in the secondary winding (E_i). Because the secondary circuit is a series resonant LC circuit, the voltages developed across the inductor and capacitor are greater and differ in phase with the induced voltage. The voltage across the capacitor is actually the output voltage and it is 90° ahead of the primary voltage at resonance

resonant transformer vectors

It was repeated on the previous page, as you learned in Volume 3, that the transformer primary and secondary voltages are 180 degrees out of phase. But at resonance, this is not completely true because the voltage induced in the secondary is not the voltage across the coil. The voltage *induced* in the coil can be considered as an *applied* voltage to a series LC resonant circuit, so that the voltages *developed* across the coil and capacitor are much greater than that induced in the coil. The output of the transformer is actually the voltage developed across the capacitor, which has a different phase angle than the induced voltage. Let's work it out.

In the primary tank circuit, the current in the coil (I_p) *lags* the voltage across the coil (E_p) because the current is *inductive*. The primary current then induces a voltage in the secondary circuit (E_i) that is 180 degrees out of phase with the primary voltage, just as in a regular transformer. But since this is a tuned circuit, the induced voltage is applied as a source voltage to the inductor and capacitor in series. This causes a current to flow in the secondary circuit (I_s). But at resonance, the circuit is *resistive*, so the secondary current (I_s) is in phase with the induced voltage (E_i).

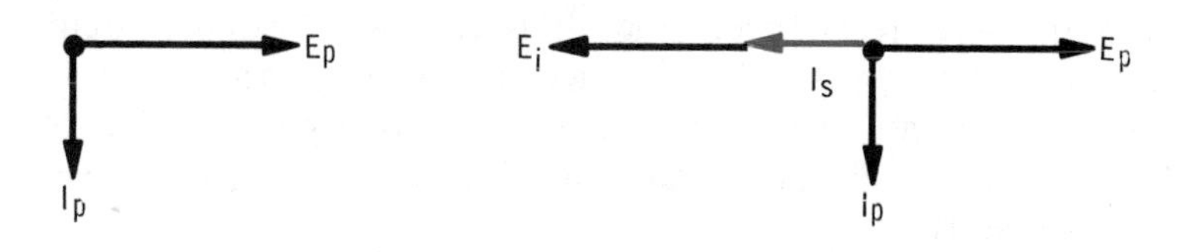

resonant transformer vectors (cont.)

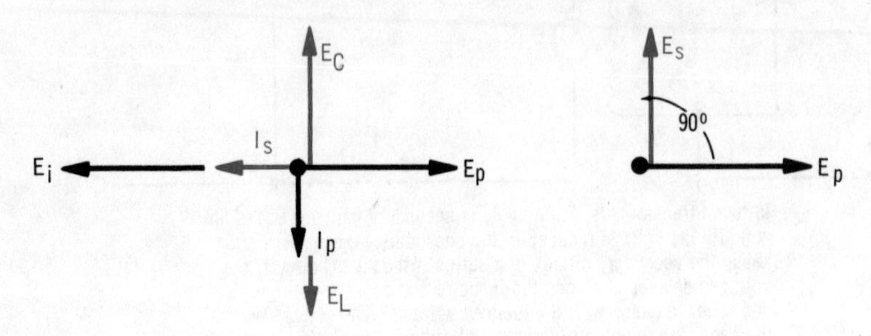

Voltages are then dropped across the inductor and capacitor, but because of the phase relationships of current and voltage in these components, the drop across the inductor (E_L) *leads* the current (I_s) by 90 degrees, and the drop across the capacitor (E_C) *lags* the current (I_s) by 90 degrees; this is the same as in an ordinary series LC circuit. The capacitor voltage (E_C), then, also lags the induced voltage (E_i) by 90 degrees. And since the induced voltage (E_i) and the primary voltage (E_p) are 180 degrees apart, the capacitor voltage (E_C) *leads* the primary voltage (E_p) by 90 degrees. Since the output, or secondary voltage, is actually the voltage developed across the capacitor, then the secondary voltage (E_s) leads the primary voltage (E_p) by 90 degrees in a tuned transformer at resonance.

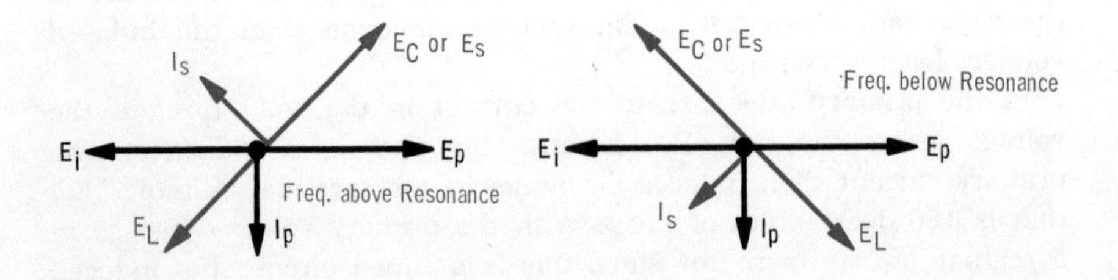

Off resonance, though, the secondary circuit is no longer resistive, so the current (I_s) will not be in phase with the induced voltage (E_i). As a result, the voltages developed across the inductor and capacitor will be more or less than 90 degrees away from E_i, depending on how far the tuning is off resonance. Since it is a series resonant circuit, the lower the frequency is off resonance, the more capacitive it becomes, and E_s will go toward 180 degrees away from E_p. For the higher frequencies, the circuit becomes inductive, and E_s will approach the same phase as E_p.

impedance matching

In an electric circuit, power is delivered by the source to the load, or loads. For the case of a simple circuit with a single resistive load, the power delivered is $P = EI = I^2R$, where R is the resistance of the load. The power, therefore, depends on the current and the resistance of the load. Since the current increases when the load resistance decreases, you may think that the smaller the load resistance was made, the larger the power would be. This would be the case, were it not for the internal resistance of the source. The source resistance is in series with the load resistance, so it also affects the current, and thus the power delivered to the load. The combined effect of the load and source resistances is such that maximum power is delivered to the load when the load impedance equals the source impedance. When the two impedances are equal, they are said to be matched. As you will find later in your electrical work, very often a low-impedance source must be matched to a high-impedance load, or vice versa. This could be done by using a source having the desired impedance, but generally this is impractical. A very common way to match the impedance of a load to the impedance of the source is by using a transformer. You will recall from Volume 3 that the impedance ratio between the primary and secondary of a transformer depends on the turns ratio of the transformers, according to the equation:

$$\sqrt{Z_p/Z_s} = N_p/N_s$$

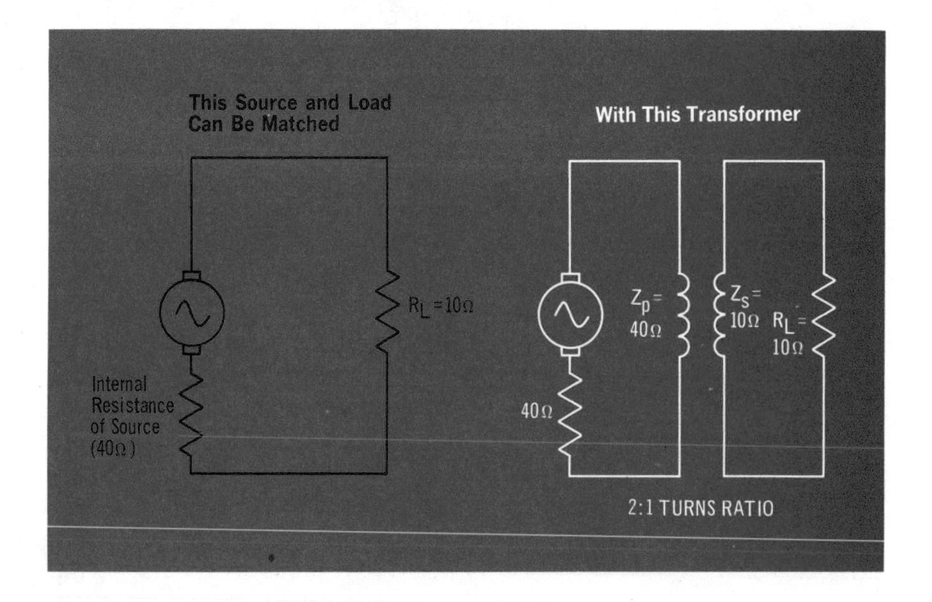

impedance matching (cont.)

Therefore, by using a transformer with the appropriate turns ratio, you can obtain any desired values of primary and secondary impedances. If the transformer primary is connected to the source, it serves as the load for the source, and if it has the same impedance as the source, maximum power is transferred to the primary. Similarly, if the secondary is connected to the load, it serves as the source for the load. And if the secondary impedance equals the load impedance, maximum power is transferred to the load.

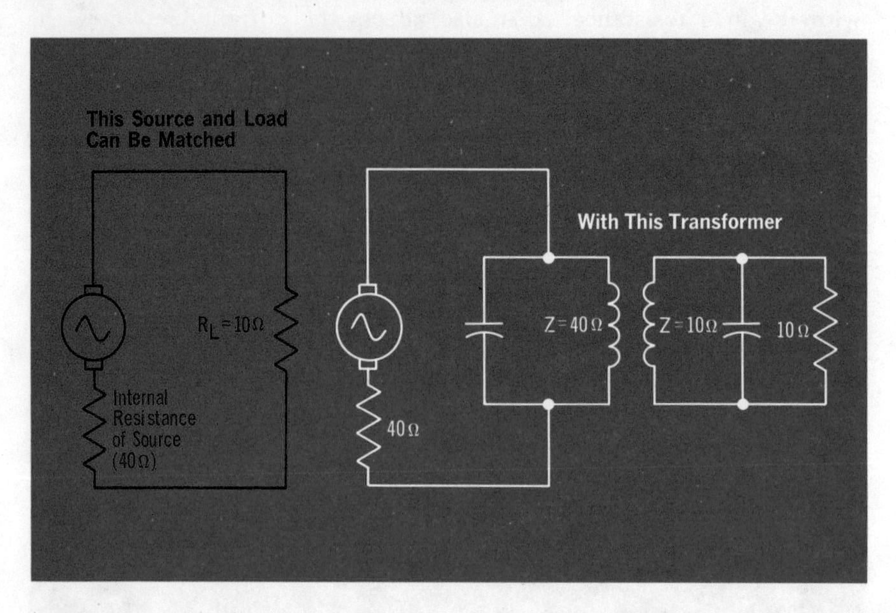

You can see that transformers are useful for matching impedances between the source and the load. In addition, transformers can step up or step down the source voltage to the value the load needs. However, since both impedance and voltage matching depend to a great extent on the turns ratio, the transformer is often designed to *compromise.* Quite often, though, better matching can be obtained if *tuned* or *resonant transformers* are used. With such transformers, the *primary* winding is usually part of a *parallel* tank circuit, and the secondary winding is a *series* resonant circuit. The turns ratio can be set for the best voltage match, since the proper impedances of the resonant circuits can be set by controlling the Q's of the coils. In addition to this, since the *primary* winding is part of a parallel resonant circuit, it has a *high* tank *current* for better inductive coupling. And the *secondary* winding is part of a series resonant circuit, so it produces a *voltage gain.*

tuned transformers

In many high-frequency applications, the tuning of resonant transformers is handled in particular ways to control the way the circuits respond to the signal frequencies involved. Because of the amplification factors involved in the primary tank circuit, the transformer coupling, and the secondary tank circuit, the bandpass action of the entire tuned transformer can provide a variety of bandwidth curves, depending on how the transformers are tuned and coupled.

When both tank sides of the transformer are tuned to the same frequency, the bandwidth of the circuit can have a sharply sloping curve that is moderately flat at the top as long as both windings are properly coupled. The exact width of the curve will depend on the Q's of both resonant circuits. If the transformer is *overcoupled*, the passband will widen and the top of the curve will dip in proportion to the amount of overcoupling. *Undercoupling*, on the other hand, will reduce the height of the curve as well as its rate of slope.

If the primary and secondary resonant circuits are tuned above and below the desired center frequency, bandpass widening can be controlled to give more bandwidth with sharp sloping curves. This is called *stagger-tuning*. As the primary and secondary tuning go further apart, the height of the curve will drop and the top of the curve will dip, with peaks showing at the primary frequency, f_p, and the secondary frequency, f_s. If only the primary *or* the secondary is tuned off the center frequency, f_c, the bandwidth curve will skew in that direction.

Combinations of off-tuning and degrees of coupling can produce varieties of bandwidth shapes.

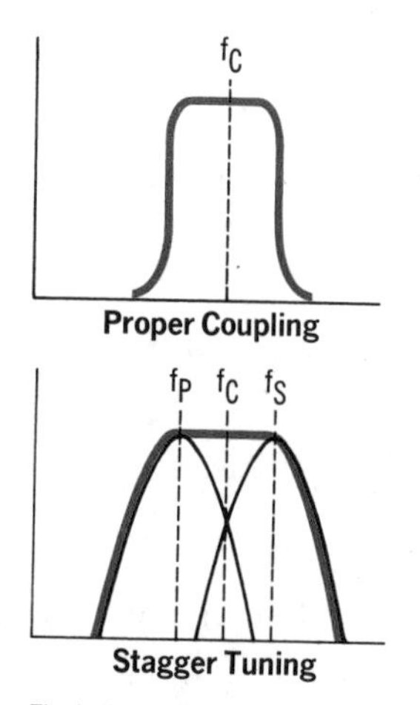

The individual tank curves combine to form one flat curve

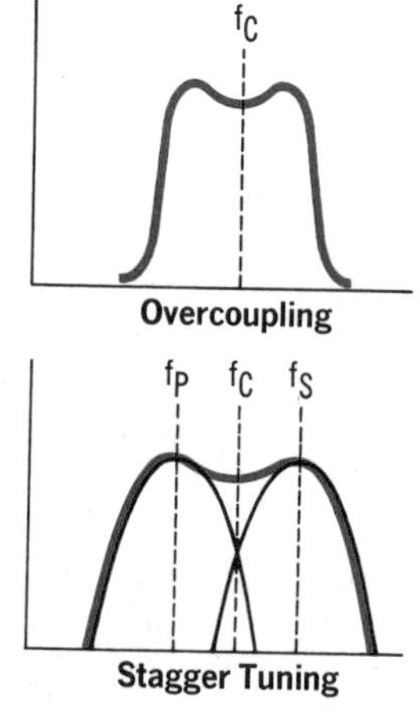

Tuning the primary and secondary further apart dips the curve

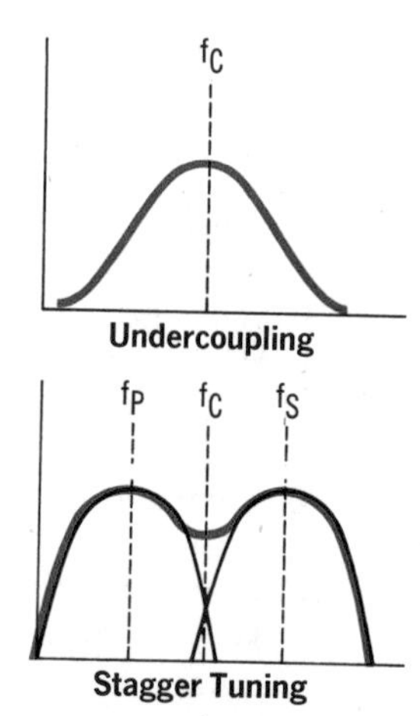

Setting f_P and f_S further apart will widen the curve and increase the dip

summary

☐ Series-parallel RL, RC, and LCR circuits contain both series and parallel combinations of resistance, capacitance, and inductance. ☐ When solving series-parallel circuits, vector additions of voltages, currents, and impedances usually have to be performed by first resolving the various quantities into their components. ☐ Tuned circuits are resonant circuits whose resonant frequency can be changed. This is accomplished by making either the capacitor or inductor in the circuit variable.

☐ Filter circuits reject voltages or currents of certain frequencies, while passing voltages or currents of other frequencies. ☐ Lowpass filters pass all frequencies up to a certain point, and reject all higher frequencies. ☐ Highpass filters reject all frequencies up to a certain point, and pass all higher frequencies. ☐ Other types of filters are bandpass, and band-reject filters. ☐ Transformer voltages and currents can be represented by vectors. When this is done, the primary voltage is used as the phase reference. ☐ In resonant transformers, the primary and secondary voltages are not 180 degrees out of phase. Instead, the secondary voltage leads the primary voltage by 90 degrees at resonance.

☐ Transformers are often used to match the impedance of a source to that of a load. Such impedance matching is accomplished by using a transformer having the appropriate turns ratio. ☐ The impedance ratio between a transformer primary and secondary depends on the transformer turns ratio according to the equation: $\sqrt{Z_p/Z_s} = N_p/N_s$.

review questions

1. What is a *series-parallel RL, RC,* or *LCR circuit*? Draw a a schematic diagram of one such circuit.
2. Draw another type of series-parallel RL, RC, or LCR circuit.
3. What is a *tuned circuit*? Draw a schematic diagram of such a circuit.
4. Why are tuned circuits used?
5. What is a *filter circuit*?
6. Draw the frequency response curve of a band-reject filter. Of a low-pass filter.
7. Describe a simple low-pass filter and its operation.
8. A capacitor connected in series with a load is what kind of filter?
9. Draw a vector diagram for a typical transformer.
10. What is meant by *impedance matching*? Why is it important?

index

index